# Business Policies in the Making

By the same author

*The Rise and Decline of Small Firms*
 (George Allen & Unwin, 1973)
*Social and Business Enterprises*
 (George Allen & Unwin, 1976)

# Business Policies in the Making

## Three Steel Companies Compared

Jonathan S. Boswell

*Senior Lecturer in Managerial Economics*
*City University Business School*

London
GEORGE ALLEN & UNWIN
Boston          Sydney

338.7669142
B74 b

**George Allen & Unwin (Publishers) Ltd,**
**40 Museum Street, London WC1A 1LU, UK**

George Allen & Unwin (Publishers) Ltd,
Park Lane, Hemel Hempstead, Herts HP2 4TE, UK

Allen & Unwin Inc.,
9 Winchester Terrace, Winchester, Mass 01890, USA

George Allen & Unwin Australia Pty Ltd,
8 Napier Street, North Sydney, NSW 2060, Australia


First published in 1983

**British Library Cataloguing in Publication Data**

Boswell, Jonathan S.
  Business policies in the making.
1. Industrial management – Great Britain – Decision-making
I. Title
658.4'012'0941      HD70.G7
ISBN 0-04-330335-8


**Library of Congress Cataloging in Publication Data**

Boswell, Jonathan.
  Business policies in the making.
Bibliography: p.
Includes index.
1. Steel industry and trade – Great Britain. 2. Dorman Long (Firm)
3. United Steel Companies. 4. Stewarts and Lloyds.    I. Title.
HD9521.5.B64   1983      338.7'669142'0941      83-2669
ISBN 0-04-330335-8


Set in 10 on 11 point Century
by Computape (Pickering) Ltd, Pickering
and printed in Great Britain
by Biddles Ltd, Guildford, Surrey

# Contents

# Foreword

This is a historical study of business policies within the context of theories of management and the firm. It addresses three core themes: (1) degrees of rationality and error in decision-making; (2) trade-offs between growth, efficiency and social action; and (3) managerial specialisation and phases in the development of firms. These themes are illustrated and tested against the experience of three large iron and steel companies between 1914 and 1939: Dorman Long, the United Steel Companies, and Stewarts and Lloyds.

The book is mainly intended for people interested in management: practitioners, students, educators and theorists. It should also be of interest to historians of business and of the interwar period, and to those concerned with the iron and steel industry.

This study is partly a development of two previous ones. In *The Rise and Decline of Small Firms* (1973) I analysed the managerial and economic stages of development through which such firms appear to pass, using both field studies and demographic data. In *Social and Business Enterprises* (1976) I discussed how basic economic concepts can be used to illuminate management decision-making. Both these strands are continued here.

I became convinced, however, that such issues, particularly as related to large companies, demanded a historical view. It seemed unlikely that recent developments in any large business would provide adequate material because of insufficiencies in or restricted access to vital documents, understandable sensitivities among key decision-makers who were still alive, and the difficulties of observing the complex origins, let alone the full working through, of important decisions. At the same time, rich documentary evidence was becoming available on the prewar history of some large firms in a particularly important industry, iron and steel. I felt strongly that the tortured postwar convolutions of nationalisation, denationalisation and renationalisation, should not be allowed to obscure that industry's power to illustrate wider managerial phenomena. I believe, too, that the continuities of management problems and corporate patterns are sufficiently strong for these prewar sagas to have much contemporary relevance: all the more so when several large firms are surveyed analytically and comparatively.

The book is arranged as follows. Chapter 1 introduces the core themes and hypotheses, while Chapter 2 briefly sets these in the context of the British iron and steel industry. The historical analysis follows in Chapters 3 to 9. The three companies are introduced in Chapter 3, which compares their growth policies in the period 1914–20. Responses to economic adversity in the early 1920s, with particular reference to comparative efficiency pursuits, are discussed in Chapter 4. Critical changes in top management between 1925 and 1931 are treated in Chapter 5. The remaining historical chapters largely concentrate on the 1930s: successes and failures in growth in Chapter 6; efficiency and organisational problems in Chapter 7; relationships with public opinion and government (including some aspects from earlier years) in Chapters 8 and 9. At each stage the historical material is integrated with the core themes and theoretical issues. Finally, Chapter 10 discusses the conclusions and their implications for ideas about management and the development of firms.

The book results from work that I have undertaken on and off over four years, mainly in the British Steel Corporation archives, and also in the Public Records Office, the Bank of England, and other places, to which have been added a secondary literature survey and interviews with a number of people who kindly provided recollections of the industry, the firms and their leading personalities.

Grateful thanks are due to The City University for granting me a sabbatical, which was largely spent on finishing the book, and to the Nuffield Foundation for financial assistance.

I owe a tremendous debt to many people, too numerous to mention, with whom I have discussed the issues over many years spent first in industry, then in academic life. Particular acknowledgement is due to Professor Alfred D. Chandler, Jr, for his encouragement and stimulus in connection with some earlier papers, when I was working as a Research Fellow at the Harvard University Business School in 1979.

I am grateful to Professor G. C. Allen, Gordon Boyce, Professor Elizabeth Brunner, Dick Edwards, Rob Grant, Professor Sir Douglas Hague, Dr Leslie Hannah, Professor Sidney Pollard, Professor Leslie Pressnell, Professor Barry Supple, Professor Anthony Slaven, Dr Steven Tolliday and Dr Kenneth Warren for helpful advice on relevant issues at various times. To the following thanks are owed for the provision of important documentary materials: Dr Jake Almond, Charles Blick, Professor Elizabeth Brunner, Jack Lancaster, E. G. Saunders and Dr Clemens Würm.

Valuable assistance over archival sources was given at the Bank of England by the late Eric Kelly and by John Keyworth. At the British Steel Corporation my debts are particularly extensive. I should like cordially to thank Derek Charman, Peter

Emmerson and Jill Hampson and, among the people who helped so courteously in the BSC's regional records centres, Elizabeth Kitson, Carl Newton and Terry Whitehead.

Grateful thanks are due to the following for agreeing to be interviewed and giving me the benefit of their recollections: Philip Beynon, R. S. H. Capes, Mrs Grizel Castle (née Macdiarmid), Sir Monty Finniston, Tim Forder, George Gowland, the late Stephen Gray, Mrs Elspeth Holderness (née Macdiarmid), James Jack, Edward T. Judge, David Ward Jones, Tom Kilpatrick, Mrs Rachel Kydd (née Benton Jones), Sir Peter Matthews, Richard Ottley, Colin Over, R. M. Peddie, J. E. Peech, Robert M. Scholey, Professor Sir Robert Shone, J. G. Stewart, Mrs Agnes Tubman (née Lyons), and W. N. Menzies-Wilson. And to E. G. Saunders and the late Douglas Wattleworth for written answers to questions.

Acknowledgement is due to the Editors of *Business History* for permission to use material from an article published in that journal in January 1980.

I am grateful to Edward T. Judge, Professor Sir Robert Shone and Dr Kenneth Warren for their helpful comments on earlier drafts. Needless to say, neither they nor any of the other people mentioned above are to be associated with the judgements and arguments of the book, for which I take full responsibility.

Finally, I should like to express gratitude to Dr Bruce Johns for valuable help in some archival work overlapping with another project, where he was my research assistant in 1978–9; to Mrs June Wates for assistance with newspaper sources; and to my wife and family for much understanding and patience.

Jonathan S. Boswell
August 1982

# 1  Management: trade-offs, biases and phases

Few important groups in modern society have neglected their own history as much as managers. Soldiers, doctors, lawyers, teachers, scientists, politicians, trade unionists – all appear to have shown far more interest in the historical development of their professions. They have looked to the past, partly to help in understanding how their arts and skills came to be what they are, partly because of an acceptance that past experience still has many lessons for policy today. By contrast, business management's attitudes have been remarkably ahistorical. Moreover, this lack of concern is shared by those who should be better able to stand back from the current hurly-burly in the cause of a long perspective – the management educators. The neglect of management history is all the more surprising because of the avalanche of literature about management problems and techniques which has poured forth over the past two decades or more.

The consequences are serious first of all for management itself. Much that could contribute to a proper sense of collective dignity and pride lies buried. Ignorance of the economic and political tribulations confronted by their predecessors encourages a belief that contemporary adversities are special and, perhaps, also encourages a tendency to over-react. Managerial understanding of an industrial decline like Britain's is distorted when the deepseated origins of that century-long process are ignored. Managerial gloom over the contemporary difficulties of the Western advanced economies generally is compounded when the realities of past depressions – let alone the achievements of past managements in confronting them – are neglected. Further, an absorption in the 'recent' and the 'current', torn from their historical context, may also produce complacency. It is this that underlies the optimistic assumption that there has been a great collective advance or even a revolution in managerial techniques and capacities. This view has no solid empirical support and remains no more than an interesting hypothesis.

For outside observers of management even greater distortions arise. Contemporary studies arc unlikely to uncover either long-term trends or managerial conflicts and errors. Naturally enough, companies resist searching enquiries into boardroom arguments, failures of judgement, or alternative strategies considered and refused. Consequently, in the absence of proper historical correctives, all manner of superficial pictures of management continue to flourish, often fed by one-sided journalistic exposés or ideological biases: conspiracy theories; ascriptions of super-efficiency (socially benign or lethal, depending on one's point of view); notions of a rat race or a soulless technocracy; assumptions unconsciously inherited from Keynes's 'defunct economists', notably Adam Smith and Karl Marx.

This neglect of management history has equally baneful consequences for academic work on business behaviour. It helps to make much theoretical work in this field repetitive and esoteric. It contributes to the assumption, tempting for adherents of some familiar 'theories of the firm' but signally unwarranted, that business behaviour has already been largely explained. It helps to preserve, for example, the continued influence of the profit-maximising hypothesis which, whatever its merits in predicting competitive market behaviour, has severe limitations when it comes to understanding the nature of management and decision-making in firms. The management history gap makes many important issues much harder to investigate, notably patterns of company success and failure over long periods, the relative importance of economic forces and strong individuals, and phases in the long-run development of firms.

If we want an idea of what really happened, then only one effective method exists: a detailed history covering a long period and written well after the events concerned. The question may be asked, why choose 1914–39 as the period of study? I was influenced in my selection by three main considerations. First, a period of such length is needed because economic ups and downs affect the evaluation of business failure and success. Major decisions, for example over capital investment, mergers and diversification, often take fifteen or twenty years to bear fruit, whether for good or ill, generally through several turns in the economic cycle. The 1914–39 period, although largely characterised by recession, contains a sufficient variety of economic conditions to allow for fair judgements of performance.

Secondly, avoiding a recent period reduces the problems of corporate managerial bias. A capable contemporary historian, interviewing managers in a large, well-known firm, may doubtless produce colourful topical material. But almost certainly such interviews will be swamped by immediate crises and decisions, and flawed by the public relations rationalisations, the romanti-cisations and self-justifications to which leading businessmen, like the rest of us, are prone. Even after an interval of fifteen or twenty years a successor management may show biases: a pro-tegé's loyalty to his ex-boss, sensitivity towards the feelings of predecessors who are still alive, sometimes a contrary tendency to denigrate predecessors in order to show the new management in a better light. There is often, too, a nagging doubt as to whether those agreeing to co-operate are representative. Such problems can be overcome by surveys of large representative samples in which anonymity is provided. But these impose an unwelcome sacrifice of identifiability as well as of length of perspective.

A third reason towers over the rest. Only in this way, almost invariably, can comprehensive documentary evidence be obtained. Even sophisticated and co-operative firms are reluctant to give a historian full access to, and use of, their relevant papers, except for periods long past. Often their history makes full sense only if the papers of some other companies are examined, perhaps those of near competitors, and also those of the relevant trade associations or national bodies. Such additional material is even less likely to be available for a recent period. Moreover, the political and social aspects of management decisions often sug-gest that official records should be consulted but in Britain these are subject to a 'thirty-year rule'. Normally, therefore, only a thirty- or forty-year wait will ensure the necessary access to managerial intimacies, boardroom secrets, external relation-ships, and details of how the firm looked to others.

This study is a comparative one, analysing the 1914–39 man-agerial experience of three companies, Dorman Long, the United Steel Companies (USC), and Stewarts and Lloyds. The sources on these companies are particularly full. All were large and impor-tant firms which contended with a wide range of problems. They displayed significant contrasts in strategy, attitudes and per-formance, and they all went through managerial changes during the period. The comparative approach enables their managerial characteristics, successes and failures to be treated more fairly than is possible in a single company history. Such comparability has been feasible because, following nationalisation in 1967, the British Steel Corporation organised the records of the previously independent companies, and because that organisation observes the thirty-year rule. The BSC factor also ensured the availability of the records of other firms with which our three had important dealings, together with those of the industry's national federation, which played a significant role. If the industry had remained in the private sector, it is possible that one of the firms would have co-operated but unlikely that all three would have done so, let alone the other diverse interests involved. Of course, the thirty-year rule means that the present study could have gone up to the late 1940s. But 1939 was preferred as the cut-off point because it is arguable that conditions during the Second World War and its immediate aftermath were exceptional compared with the periods both before and since.

A warning to the reader is necessary. Most company studies provide an exhaustive description of the firm, its products, markets and innovations, its strategies, organisation and personnel. They are rich in circumstantial detail, ranging from incorporation dates through technical developments to works outputs. Despite the great value of such exercises, that path will not be followed here. A comprehensive approach to the history of three large firms over twenty-five years would make the book far too long; in fact, three full-length books would be needed. More important, this study is deliberately selective. It assumes that history's usefulness rests considerably on its ability to stimulate thinking and test hypotheses, and that it should somehow be combined with theory. Such a combination is particularly desirable in management studies. For here empirical history and theories mainly from the social sciences have long continued to neglect each other with the consequence that both are impoverished.

Concentrating on a few central issues involves fewer sacrifices than might be supposed. Of course, readers who desire much technical detail will be disappointed, although it is hoped that the full footnotes provided at the end of each chapter will guide specialists towards the materials they seek. It should be strongly emphasised that a thematic concentration does not imply a loss of objectivity. The writer's conscience should mean that wherever the data do not fit some guiding idea or theory, this is admitted, and every effort has been made to attain such objectivity here. However, the notion that objectivity is possible in the sense of a complete detachment from theories and values is, of course, an illusion. Even the historian most devoted to the notion that 'the facts speak for themselves' necessarily uses theories and values in selecting data and explaining 'what happened'. This is all the more the case if he seeks to explain why things happened as they did, let alone to interpret their significance. The difference between his approach and that of the present study is, therefore, only a matter of degree. But there are also positive advantages in thematic concentration. For without guiding themes and relationships with theory it is easy to slip into formless description: easy to miss the perennial issues of management which make its history relevant to the policy problems still being faced today.

A final introductory point is that this book is concerned with business policy, that is to say the central objectives, concepts and decisions of the firm. This includes, very importantly, what is usually called 'business strategy', namely the firm's responses to market opportunities and problems in terms of specialisation and diversification, horizontal and vertical integration, mergers, pricing and other forms of competition, and so on. But business policy includes much more than strategy in this sense. It stretches upwards to embrace the firm's objectives, values and purposes, and sideways to cover its organisation, notably its policies on centralisation and decentralisation. Business policy also includes the neglected field of relationships with public opinion, social issues and government. In fact, it is comprehensive and synoptic. It is also para-functional, transcending the conventional divisions of marketing, production, finance, etc. Its formulation constitutes what is, perhaps, the most distinctive task of management, using this term to signify the top direction of the firm.

The rest of this chapter briefly outlines the three sets of themes relating to business policy with which the book is concerned. These themes relate to decision-making processes, managerial or corporate objectives, and long-term phases in the development of firms.

## Decision-making processes: rationality and error

To begin with, there are the classic issues of how business policy is formulated and applied, of the methodology of decision-making. How explicit was the formulation of policy? How comprehensive was the information mobilised for decision-making? How conceptually clear was the use of that information? How fully were the diverse abilities of management employed in reaching key decisions? What major errors occurred, both of commission and omission, and how far could these have been avoided?

This field has been dominated by three main approaches.[1] The first may be termed broadly the rationalistic view. According to this view, management pursues clear and consistent objectives (just a few or only one), considers all the available alternatives, calculates or estimates the costs and benefits of each of these, and then decides on the 'optimal' course of action. The most rigorous form of this approach, which derives from microeconomic theory, sees decision-making as a matter of quantified calculation. It assumes that the firm: (a) pursues a single, constant, quantifiable objective, usually maximum profit; (b) observes various quantifiable constraints; (c) clearly perceives a full range of alternatives; (d) accurately forecasts future trends or at least assigns numerical probabilities to them; and, therefore, (e) arrives precisely at an 'optimal' decision. Technically, profits are maximised when marginal costs and marginal revenues are equal or, in the case of diverse activities, when the marginal returns from applying identical inputs to those activities are also equal. This is the theory at its purest. However even in its less rigorous forms, as set out in many textbooks, the rationalistic view adheres to the same spirit. It lavishly imputes logic, unity and consistency of purpose, clarity and predictive power to management or at least holds that these desiderata are attainable.

At the other extreme is the approach often described as 'behavioural'. According to this view, the rationalistic theory just outlined is unrealistic. A better hypothesis is, for example, that 'administrative policy characteristically involves a continuation of past policies with the least possible, i.e. incremental, modification to suit changing circumstances'. Decision-makers 'make do' with decisions that 'get by', the easiest being those that follow past precedent. Management processes involve searching for a 'coalition' or consensus among groups inside the firm which are separate or even conflicting, not homogeneous and united. Decision-making involves 'satisficing', or the pursuit of minimal or acceptable standards, rather than 'optimising'; balancing efforts or 'appreciation' rather than weighing-up quantified alternatives; a search for 'better' solutions rather than the 'best'. Such phenomena, it is sometimes suggested, are not necessarily irrational. Indeed, they may attain the status of 'a science of muddling through' and may be both effective and creative from the viewpoints of the firm and society. They are definitely felt by their protagonists to be good descriptions of what managers actually do.

Although both of these approaches are useful as conceptual devices, in my view they are too extreme. If the first is too elegant and formalistic, the second is merely formless and loose. Fortunately, there is a middle position, one which concentrates on degrees of rationality in decision-making.[2] This approach makes various concessions to the 'behavioural' view: in particular, it says that risk and uncertainty do restrict managers' forecasting abilities, that the ordinary human constraints of intellect and imagination do limit their powers of comprehension, and that decision-making is obviously further complicated by conflicts of interests and values inside the firm. Depending on these various constraints, however, the rationalistic ideal can be, and is, approximated to in many specific ways. For example, although pure quantified problem-solving is inapplicable to major strategic decisions, it is perfectly appropriate for various lower-level ones. Although firms clearly lack a precise knowledge of their marginal costs and revenues, they may be perfectly capable of estimating the 'marginal' effects on profits of making particular decisions, for example on output, investment or prices: perfectly capable, too, of estimating the contribution of particular activities to profits. It is possible for managers to make an implicit use of such concepts as discounting and the opportunity cost of capital, to marshal several alternatives, to estimate the costs and benefits of each, including the alternative of 'staying put', to employ probability thinking by, for example, preferring ranges of estimates to single-figure forecasts. A top management can think imaginatively about competitor reactions in an oligopolistic market, it can ensure that the firm's best minds are mobilised in its major decisions and it can use deliberation and reasoned argument in arriving at such decisions.

The implied hypothesis is that such approximations to rationality have always been practised, at least by some. The tendency for semantics to change and for modern management techniques to attain higher levels of specification and apparent precision does not alter the likelihood that during any period these fundamentals have always existed in some firms in varying mixtures and degrees. But if a primary task of management history is to investigate varying degrees of rationality in this sense, several things are required. It may be necessary to go beyond the precise evidence and to apply wider knowledge about managerial behaviour. It is certainly necessary to study a wide variety of data. For example, much depends on what is unsaid. Many of the relevant materials are disconcertingly widely spread about, blatantly disregarding any rubrics of subject matter, managerial hierarchy or chronology. Almost any source can bear on almost any aspect of policy. Major treasures may be buried in a minor sub-committee's minutes, a pencilled scrawl in a margin, a casual aside, a holiday letter from some seaside resort – a trivial point suddenly illuminating some major dispute.

Above all, the historical study of degrees of rationality in business decision-making has to grapple with the question of alternatives, including the thorny question of 'errors'. Misjudgements should be related to the issue of what *else* a firm could have done. 'Errors' about the future should be classified according to whether they related to the macro economy, to sectors outside the firm's experience or to its own market. Account must be taken of the information available to decision-makers at the time and of the other courses of action which they considered or which, arguably, they should and could have considered. The study of business policies has suffered from a narrow type of *ex post* view in which, ironically, the opportunity costs which those policies necessarily involved for the firm or society are themselves lost. Business events are described as if they had to happen. There is a slide into implicit determinism, even into a sort of Whig business history in which what happened also appears to have been 'for the best'. Hence business history has a long way to go in catching up with political history, economic history or biography, where rich and stimulating controversies have long proliferated about not only the why's but also the might-have-beens.

A first objective of this study, then, is to examine degrees of rationality and error in decision-making, exploring the middle ground between the rationalistic and 'behavioural' theories whilst using these extreme views as conceptual benchmarks. For this purpose it concentrates on successive policy challenges of the period when the decision-making rationality of the three firms simultaneously faced comparable tests. These challenges emerged, broadly, in the economic and psychological conditions of war and its immediate aftermath between 1914 and 1920; in the recession years of the 1920s; in top management changes between 1926 and 1931; and, by the 1930s, in major expansion projects and internal rationalisation efforts.

But these same phases illuminate more than comparative methods of decision-making. They can also shed light on the firms' overarching values and purposes. This brings us to a theme which has long been at the storm centre of discussion of theories of the firm, business behaviour and management; even longer: the question of the firm's objectives.

## Growth, efficiency and social action

This study suggests that business policies crystallised round three dominant pursuits: growth, efficiency and social action. Once again this represents an intermediate view between two extremes. On the one hand, there is the familiar hypothesis that the firm tries to, or behaves as if it does, pursue a single objective, as in 'economic theories of the firm' which posit a single maximand, usually profit. On the other hand, there is the assertion that firms' objectives are utterly variable, pluralistic and shifting. If the former approach is basically monist, the latter suggests heterogeneity and indeterminacy. In what follows both extremes are rejected in favour of an approach similar to that often used in analysing national economic policies. The idea is that we can best make sense of business policies by assuming a few objectives between which there are trade-offs.

It should be said at once that the wider uses of profit-maximising theory are not disputed, let alone the operational importance of profit in business decisions. The profit-maximising hypothesis helps to explain and predict the general behaviour of markets and it is a most useful conceptual tool for economists. The central operational role of profit for managers hardly needs emphasising. Obviously, it is fundamental to the whole mechanism of business enterprise. Moreover, profit relates importantly to the analysis of this book. For example, how far firms were aware of profit opportunities is vital in evaluating degrees of rationality as just mentioned. Also, profit was a precondition and essential instrument for the wider objectives about to be outlined. But this does not necessarily mean that profit has been the central driving force leading management up and on. The suggestion here is that growth, efficiency and social action *have* been such forces; that it is these themes which have commanded intellectual, emotional and moral adherence as causes or working ideals; but also that they have formed distinct and partially conflicting poles of attraction.

Of the three pursuits, *growth* is perhaps the most simple and familiar. It has a central place in many theories: as the governing principle of the most successful parts of the business system (Schumpeter, Penrose), the embodiment of entrepreneurship and risk-taking (Schumpeter, Knight), the dominant rationale of managers in a corporate economy (Galbraith), even the single objective which managers seek to maximise subject to constraints (Baumol, Marris).[3] In this book 'growth' is taken to mean not only an output increase within a firm's existing capacity (as in the economist's 'short-run') nor simply an increase in capacity in line with other firms (as in a general investment boom). It is mainly equated with an increase in the firm's size and competitive strength relative to others in the industry. In this sense growth requires special push and effort. It reflects a rich variety of managerial motives: for example, the pursuit of money, power, market stability, outlets for under-utilised resources, employment creation, social prestige, delight in the creation of new products or consumer wants, the sheer élan of conquest, risk-taking, empire-building or winning the game.

The second dominant pursuit, *efficiency*, means trying to dispose existing resources to the most productive effect. It involves improving an input–output ratio: by cost-cutting, by increasing productivity or, in a more sophisticated sense, by so juggling diverse activities as to maximise overall returns, a process which would be fulfilled in theory at an 'optimal' point where the marginal returns from applying identical inputs to a variety of activities were equal. Thus the pursuit of efficiency applies to any function within the firm and also to its overall resource allocation, including, very importantly, its human resources, not least its management. The idea that efficiency is of comparable importance to growth as a managerial pursuit rests on two hypotheses: first, that the efficiency-organisation problems of large businesses are highly complex and exacting (see below), second that efficiency-seeking, like growth-seeking, draws on a highly diverse range of personal managerial drives: – in particular, frugality, tidy-mindedness, aesthetic instincts, the enthusiasms of accountants and engineers, the urges to dominate a group, to perfect an organisation, to draw out the best in people, and to create teamwork.

The third dominant pursuit, *social action*, reckons with the fact that managers are part of society, subject to influences from public opinion and government. The latter often suggest that managers should also follow (a) collective rather than narrowly sectional concepts of growth and efficiency, or (b) other types of 'good' behaviour like truth-telling, restraint in the use of market power, sensitivity to neighbourhood effects, economic nationalism, and service to the state. Managers may have highly diverse motives for doing such things, notably feelings of duty, patriotism or local pride, social idealism, desires for public approval or honours, pursuits of 'pay-offs' from the state, concepts of an industry's collective self-interest. Social action, then, means the voluntary managerial pursuit of any activities, apart from the firm's own growth and efficiency, sectionally conceived, which are approved by public opinion and government. Typically, it takes concrete forms in philanthropic spending; managerial time on public work; co-operative processes *vis-à-vis* other social groups or government; industry-wide and publicly-approved collective action. Social action also acts as one influence among others on commercial restraint as, for example, when a firm holds down prices partly to keep out new competitors but partly also on grounds of equity, social repute or compliance with public policies. It is not suggested that social action is on the same level as growth or efficiency-seeking, indeed it may rank considerably below them in the usual hierarchy of managerial objectives. The proposition is that social action has an independent power of attraction, yields socially significant results, and affects the firm in concrete ways, often by limiting immediate profit-taking; also that it may be so complex and dynamic as to be irreducible to a mere 'constraint'.[4]

Each of these three central pursuits – growth, efficiency and social action – is readily observable in both attitudes and behaviour. Moreover, each is complex and challenging enough for its interpretation to form a major managerial task. But the central policy problem, according to the perspective of this book, is how to perceive and resolve the trade-offs which they engender.

It is obvious that there are important affinities between the three pursuits. For example, successful investment projects help both growth and efficiency. Research suggests long-run positive correlations between growth and profitability, and also between growth and productivity-mindedness or economising attitudes. A virtuous link between growth and efficiency in the very long run is suggested by Alfred D. Chandler's broad historical canvas of the development of successful large businesses. It is idealised in Edith Penrose's theory of the 'receding managerial limit'. This sees the capable firm's abilities for growth as virtually limitless. This is because persistent slack, inefficiency and frustration within its most critical resource – management – provides the dynamism, and also the potential, for further growth, precisely in order to exploit the managerial resources more fully, in other words to make them more efficient (for objections to this theory, see below).[5] As for the affinities between social action and both efficiency and growth, these are familiar: for example, overseas ventures which are patriotic as well as expansive, mergers blessed by the state, compliance with government requests on pricing or investment as a *quid pro quo* for tariffs or subsidies which help the firm's commercial policies.

Some of the conflicts between the objectives relate to time-lags and adjustments, and also to unforeseen economic circumstances. This is particularly the case with certain familiar splits between growth and efficiency. For example, technological innovations improve productivity but may have a lagged benefit for growth. Conversely, a firm quickly expands by investing heavily or taking on extra managers to run new activities, but the economies from the former take, say, two or three years to emerge, while the advantages of the latter cannot be fully achieved until the new managers are fully absorbed into the existing team. Thus in both cases efficiency lags behind growth. Often, too, the discontinuities are worsened by the economic cycle. Thus a major investment project, aimed at improving both growth and efficiency, is stymied by a recession of unpredicted timing and severity, leaving the firm with unused capacity and a fruitless expansion, so that neither cause is served.

However, still greater tensions between growth and efficiency may spring from the very nature of the firm and of management. Economic theory has long suggested that organisations, as well as single activity plants, may grow too big to be properly efficient because of problems of management control and human communications. Such pessimism receives a further twist in the 'organisational failure' theories of Anthony Downs and Oliver Williamson. These effectively deny management's triumphant role in harmonising growth and efficiency (Penrose's virtuous circle). They hypothesise by contrast an endemic growth–efficiency conflict in the very large firm.[6]

Although such gloom appears overdrawn, the underlying arguments cannot be dismissed. At the very least the implication that problems of organisation, structure and control in large firms demand painstaking care seems correct. The notion that a 'decentralised' multi-divisional structure (itself a slippery concept) resolves the size-growth *versus* efficiency problems of large diversified firms appears implausible. Even within such a structure, conflicts seem to persist between the head office's pursuit of overall efficiency or financial optimisation and the idea of 'decentralisation' as being good for morale, personal development or operational effectiveness.[7] The extent of these problems needs to be tested against the detailed experience of the large companies investigated in this book.

So does the hypothesis of substantial conflict between social action and both the other objectives. This is the little examined field of socially or politically motivated diversions of resources to the detriment of profit. Philanthropy, market restraint or sensitivity on, say, redundancy typically impose an immediate cost, either directly or in terms of commercial opportunities foregone. Any tangible benefits to the firm's growth or efficiency are likely to be long-term, marginal and uncertain, even assuming they are hoped for. It is this field which imposes perhaps the most acute moral dilemmas. For the immediate rewards of social action are probably largely psychological for managers: as Alfred Marshall put it, for 'the approval of their own conscience and ... the esteem of others'.[8] Again, the nature of such conflicts demands careful historical probing.

All of this implies a picture of management as a continuous balancing act between the three pursuits. Such a process would be marked by strong adherence to all of them, persistent searching for their affinities, clear awareness of any sacrifices of one to the other, and quick countervailing actions to correct imbalances. Thus if a firm expanded rapidly through mergers, moves would speedily follow to improve the newly-absorbed units' efficiency. If economic crisis prevented growth through investment, increasing rather than improving production costs, management would quickly put through non-production economies. But the satisfactions of reducing waste, redeploying resources and tuning control systems to concert pitch, would not long deflect attention from, say, expansion chances abroad. Again, suppose growth–efficiency aims led to defiance of industry-wide collective interests, exacerbation of unemployment in a depressed area, or the spoiling of land. The management would make up for this either by speedy direct restitution or by another form of social action related to the original lapse.

Is this picture realistic? It assumes a continuous, high-pitch balancing act between three causes to all of which managers are universally devoted. Relative valuations of the three pursuits are viewed as uniform between decision-makers, so all that varies is skill in combining them. Sacrifices of growth, efficiency or social action, arise only because of inherent time-lags, bad luck, or failures in forecasting. The underlying assumption, shared incidentally with most existing theories of the firm, is still that managerial preferences and abilities are basically homogeneous, so that the key question is merely one of 'more or less' along a single spectrum. But supposing managers are biased from the start as between growth, efficiency and social action? Suppose also that their biases are deep-seated and continuous? Combined with the conflicts already mentioned, would not this have striking effects on the long-run development of firms?

## Managerial specialisation and corporate development

In extreme terms managerial specialisation between the three pursuits would imply three pure personality types, crudely characterised as the growth man, the efficiency man and the social activist. But since such a high degree of polarisation seems unrealistic, it is more reasonable to suppose that top managers typically unite all three inclinations but in varying mixtures. Because social action is generally less prominent than the other two pursuits (see above), it is unlikely to achieve a position of dominance over them. Therefore, the possible types of main emphasis for individual managers would be as follows, where G = growth, E = efficiency and S = social action:

$$G + E + S \qquad\qquad E + S$$
$$G + E \qquad\qquad\qquad\quad G$$

$$G + S \qquad\qquad E$$

This specialisation hypothesis rests on assumptions about both preferences and abilities. The influences on managers' ideals, values and aspirations (parents, schooling, locality, political, economic and intellectual currents during formative years, earlier training, job experience) are likely to produce varying hierarchies of preferences between growth, efficiency and social action. At the same time it is assumed that abilities are also likely to be biased, reflecting genetic factors, education and earlier experience. Similarly, managers' stances on social action may be partly influenced by the highly variable factor of previous experience of different social groups.

Part of the hypothesis is familiar. It has long been customary to identify the 'entrepreneurial type' with those motives and capacities typically associated with growthmanship rather than with the things that appear to make a good administrator, a rationaliser, a reformer. There is indeed some evidence that successful founder-entrepreneurs are not necessarily so effective in controlling the firm during its later and larger stages. Conversely, although there is less theory and evidence to this effect, it seems reasonable to suppose that individual excellence in orchestrating existing sets of resources, the prime test of efficiency, probably leaves little room in a personality for strong expansionary drives.[9] As for social action, it is probable that only a proportion of leading businessmen have large supplies of social feelings and ideals.

From the viewpoint of the whole firm, of course, managerial specialisations would not matter if they were capable of being blended in the right proportions. But this cannot be easy, to say the least. It is likely that even large businesses are often dominated by a single individual, so it is his biases which then probably prevail and set the tone. An effective top management group is usually very small, making for difficulties in balancing. And a team within which specialised strengths become complementary rather than conflicting probably demands considerable nurturing. From all of this springs the likelihood that most firms are strongly influenced by individual managerial biases. Again, these would matter less if the managers could correct their biases over time. But specialisations towards growth, efficiency or social action are probably well established by the time a man reaches the top, and are a matter of deep-seated orientation rather than mere technique or skill.

The argument of the last four paragraphs defies some widely accepted notions about management's homogeneity, collective character and flexibility. Contrary to the idea of a unitary phenomenon varying in intensity or ability along a single spectrum, it asserts the likelihood of multi-dimensional biases, albeit with the important qualification that, according to the hypothesis, these biases concentrate around a few broad categories of specialisation. As against the idea of management as a highly collective process, the argument posits the likely frequency of individual concentrations of power, particularly in the person of the chief executive. Not least important and perhaps particularly unsettling to conventional images, the idea of management as highly plastic and flexible over time is denied. Rather, it is suggested that leading managers' basic orientations on growth, efficiency and social action, affecting policy stances over a wide front, are likely to be relatively stable.

If the argument is valid, a critically important inference follows, one which equally differs from certain widely-held views of business behaviour. The inference is that whole firms would experience phases of policy concentration like those outlined above for individual managers. That is to say, firms would pass through periods dominated by growth-seeking, efficiency-seeking, a growth–efficiency balancing act, or any of these qualified by social action. On any reckoning, moreover, such periods of corporate policy bias would be fairly long. If the phases roughly correspond to chief executive tenures, they would often last for ten or twenty years: longer, therefore, than most economic cycles.

This idea of long-run policy phases, reflecting internal managerial factors, does not deny the importance of external influences. There is no suggestion that economic, market and technological forces are other than critical. In particular, it is obvious that economic fluctuations qualify the degree of success of certain kinds of policies. For example, recessions clearly make growth policies more difficult. More subtly, boom periods may well slow down, or at least alter the forms of, efficiency-seeking. Nor is it denied that varying degrees of market power and competition affect the room for manoeuvre of individual firms in the familiar ways analysed in industrial economics.

However, the implication is that these factors are far from supreme. There is a marked divergence from the view that business policies are fundamentally determined by economic fluctuations, as in the extreme image of firms 'bobbing on the economic waves'. There is a clear implication that business behaviour is far from being overwhelmingly a matter of more or less successful reflexes to the economy, the market and the technology. Instead, an interactive interpretation is preferred. But there is also a denial of the more subtle view that, within such an interaction, it is only the external influences which are systematic and measurable. If the over-simplified concept of management as more or less 'capable' or 'successful' along a single spectrum is rejected, so is the relegation of management to the status of a doubtless important but nonetheless thoroughly messy and idiosyncratic residual. On the contrary, the assertion is that managerial influences are not only powerful but also reasonably classifiable. If the idea of managerial specialisation and long-term corporate policy phases is valid, it should help us to view management's causal role more clearly and systematically.

One final question arises. Are corporate policy phases likely to follow any patterns over the *very* long term, namely periods covering several top management régimes? It is tempting to resort to Alfred Marshall's potent image of the 'trees in the forest' with its intimation of a single protracted curve of growth, stagnation and eventual mortality. But whatever their other uses, such organic analogies provide little help in this context.[10] It seems more appropriate to search for possibilities of a cyclical nature, to consider whether dialectical or repetitive processes may be at work. In particular, is it possible that in the very long run most companies are subject to sequences of alternation between policy phases of growth and efficiency? Is it reasonable to picture the main corporate trend as a process of swinging over very long periods from one bias to the other, with experiences of synthesis or balance correspondingly rare?

One obstacle in the way of envisaging very long run patterns is the apparently random factor of managerial breakdown or overall inadequacy. No theory can exclude the possibility of hiatus phases during which strong managerial pursuits of growth, efficiency or social action are lacking altogether. Indeed, lapses into such managerial indeterminacy, often involving policy deadlock or mere cliff-hanging, are fairly readily observable, but the obvious difficulty is to predict why and how often they occur. It is, perhaps, equally if not more difficult to predict the timing of strong propensities towards social action relative to the growth and efficiency concentrations. All too little is known about the roots of the relevant social inclinations among managers. However, the possibility of some systematic relationships between policy phases should not be rejected.

In fact, it is possible to glimpse a number of corporate tendencies which would influence the sequencing at least of growth and efficiency pursuits. Thus supposing a firm successfully unites growth and efficiency in one phase, a winning combination commercially, would it not try to repeat this trick in the next phase? By contrast, is it not likely that a phase of concentration on either growth or efficiency would lead to diminishing returns and an eventual reaction towards the neglected pursuit? In the former case corporate efforts would be geared towards maintaining or improving a growth–efficiency balance, in the latter case one corporate extreme would beget another. However, an important qualification arises. Given the wide extent of managerial specialisation predicated above, the two tendencies would not be equally common. An alternating, cyclical type of sequence ($G{\rightarrow}E{\rightarrow}G{\rightarrow}E$ ... ) would be more typical than a reiterative one ($G+E{\rightarrow}G+E{\rightarrow}G+E$ ... ). Crudely speaking, zig-zags would be more usual than straight lines. And the greater the individual element in policy making, with its hypothesised typical biases, the greater the likelihood of corporate development proceeding by a series of one-sided growth or efficiency thrusts rather than by a smooth progression.

On the basis of these considerations five tentative propositions emerge about very long run patterns. If two of these are apparently negative, the last three offer a glimpse of predictive value.

1 *Occasional hiatuses* under managerial régimes lacking a decided impetus in any of the three directions. For example, there may be a deadlock between growth- and efficiency-seeking elements at the top; a power struggle may obstruct determined policies of any kind; a single dominant individual may be indecisive, mediocre, or deeply conservative.
2 *The independent variability* of strong social action pursuits, their provenance probably being unrelated to that of the growth and efficiency biases.
3 *Growth–efficiency reiterations.* A capable management combines growth and efficiency (substantially the Penrose case, see above), and the resulting success is one which the firm naturally wishes to maintain. Moreover, the combination is likely to yield cumulative learning effects. Hence one combination phase is probably succeeded by another, although with varying degrees of success.
4 *Growth–efficiency alternations.* A growthmanship phase eventually means that rapidly expanded resources need to be properly absorbed and that expansionary energies are relatively exhausted. The firm is ready for a reaction. But because of chief executive specialisation and continuity this requires the departure of the growth-absorbed leadership and its replacement by an efficiency-orientated régime. Conversely, an efficiency phase eventually leads to boredom with ideas of improvement and rationalisation and a frustration of growth ambitions lower down, and also to increased internal conflicts

over jobs and power, suggesting a need for growth as a safety valve. But again the reaction is only effective when the efficiency-orientated managerial régime comes to an end through death, retirement or voluntary or enforced resignation, and is suitably replaced.

5   *The domination of zig-zags over straight lines*. Because of managerial specialisation (4) is more frequent than (3).

These various ideas about managerial specialisation and corporate development phases are undoubtedly sweeping. How far can they be tested in this study? The three firms covered in this book over a twenty-five year period threw up a total of eight managerial régimes and five changeovers of régimes (and incidentally, taking into account that some of the régimes predated 1914 or survived beyond 1939, their average length was about twenty years). This provides a basis for some limited initial testing of the hypotheses of managerial specialisation and corporate policy concentrations on growth, efficiency and social action. If three intensively-worked cases broadly support the argument, something will be gained. With regard to the further speculations about sequences of policy phases in the very long run, our aims are necessarily more modest.

## Notes

1   The voluminous literature on the three tendencies identified here is usefully introduced in, for example, C. J. Hawkins, *Theory of the Firm* (1973).
2   See Joel Dean, *Managerial Economics* (1957), also J. S. Boswell, *Social and Business Enterprises* (1976).
3   Joseph Schumpeter, *Capitalism, Socialism and Democracy* (3rd edn, 1950); Edith Penrose, *Theory of the Growth of the Firm* (1959); F. Knight, *Risk, Uncertainty and Profit* (1921); J. K. Galbraith, *The New Industrial State* (1967) and *Economics and the Public Purpose* (1974); W. J. Baumol, *Business Behaviour, Value and Growth* (1967); R. L. Marris, *The Economic Theory of Managerial Capitalism* (1964); R. L. Marris and A. Wood (eds), *The Corporate Economy* (1971).
4   For a discussion of the relevant historical evidence see J. S. Boswell, 'The informal social control of business in Britain, 1880–1939', *Business History Review* (1983).
5   A. Singh and G. Whittington, *Growth, Profitability and Valuation* (1968); Alfred D. Chandler, *The Visible Hand* (1977); Edith Penrose, *op cit*.
6   Anthony Downs, *Inside Bureaucracy* (1967). O. E. Williamson, *Markets and Hierarchies* (1976).
7   The organisational problems of some large companies between the wars are referred to in Leslie Hannah, *The Rise of the Corporate Economy* (1977). For useful evidence on some more recent cases see, for example, M. Z. Brooke and H. L. Remmers, *The Strategy of Multinational Enterprise* (1970).
8   Alfred Marshall, 'Some aspects of competition', address to Economic Science and Statistics section, British Association, Leeds (1890).

9　Well-documented companies characterised by growth-orientated founder-entrepreneurs, some of whose successors were outstanding administrators and efficiency men, include Ford, General Motors, Standard Oil and Unilever. See Alan Nevins and F. E. Hill, *Ford, the Times and the Man* (1954), *Ford, Expansion and Challenge* (1957), and *Ford, Decline and Rebirth* (1962); Alfred D. Chandler, *Strategy and Structure* (1962); Ralph Hidy *et al.*, *History of Standard Oil Company*, 3 vols. (1955–71); Charles Wilson, *Unilever*, 3 vols (1954–70).

10　Alfred Marshall, *Principles of Economics*, 1st edn (1890), pp. 375–6. But see also 2nd edn (1891, pp. 372–3, and 6th edn (1910), p. 316. For a criticism of biological theories as vacuous and/or deterministic see Edith Penrose, 'Biological analogies in the theory of the firm', *American Economic Review*, December 1952.

# 2  Management: iron and steel

There are several reasons why the British iron and steel industry between 1914 and 1939 is a good testing ground for the issues raised in the last chapter. The post-1945 convolutions of nationalisation, denationalisation and renationalisation, which appear to place it in a special and somewhat tortured category, must not be allowed to obscure the industry's power to illuminate wider phenomena of business policy and management.

First, the basic technology and economics of iron and steel pushed its leading firms to face challenging decisions, while planning already took up large amounts of resources and long periods of time. We can therefore observe decision-making about investment projects, mergers and acquisitions, specialisation, diversification, and competitive moves generally. The classic problems of degrees of rationality and error in decision-making, briefly identified in the last chapter, are thrown into suitably high relief.

Second, various characteristics of the industry meant that the trade-offs between growth, efficiency and social action were particularly complex and marked. Growth-seeking in the sense of increasing market power and getting ahead of competitors at home or abroad was a vital option, as always. The leading firms were already large, complex and diversified, thus throwing up major challenges in the field of efficiency. The industry also possessed certain marked sociopolitical features and, by the 1930s, showed an increasing involvement with public policies. This required making major choices over social action. While there were frequent affinities between growth, efficiency and social action, the conflicts between these three pursuits were especially striking.

Third, there were great possibilities for managerial specialisation in growth, efficiency and social action. Hence, in so far as firms were dominated by individuals, there was great scope, too, for long-term corporate phases reflecting such specialisation. So if there is anything in these hypotheses, it should emerge fairly clearly in a study of the industry. The industry, furthermore, was particularly vulnerable to economic fluctuations, and it therefore offers temptations to the economic determinist. But by the same token it provides a good test of the argument that managerially-derived long-run policy phases were of primary importance and persisted through economic fluctuations.

## Technological and economic factors

To begin with, it is necessary to appreciate something of the basic technological and economic features of the iron and steel industry in its historical setting in Britain. The following outline is highly simplified. The reader who wants more detail is referred to the helpful studies which already exist.[1]

A first set of parameters was set by the basic technology of the industry, particularly by its vertical character. Here one needs to appreciate the broad forms of the successive stages from raw materials to final products. First came the stage of getting the main raw materials for iron and steel: large quantities of iron ore on the one hand, and of coke on the other. The iron ores, which had to be mined and then treated in various ways, varied in quality, mainly depending on removable and non-removable elements other than the iron itself. The coke was made from coal in coke ovens. The next main stage, iron making, took place in blast furnaces where the raw iron ore was refined. This involved continuous and elaborate processes of charging, melting and reducing, using much of the coke. It produced hot liquid iron or cold pig iron, both of which products varied significantly according to how much phosphorus they contained. This was a highly variable extraneous element in the original iron ores which had not been refined out, like the others, in the blast furnaces.

It was only after these initial stages of mining, preparing the raw materials and producing iron, that the central operations of steel making could begin. In these operations the iron was refined in steel-making furnaces. Essentially, this took place as a result of oxidisation and it generally involved one of two types of process: (a) the Bessemer process, which treated molten metal in a converter; and (b) the Open Hearth process, which subjected a variety of metal types, hot, cold or scrap, to heat from outside the process. Both (a) and (b) subdivided according to whether they used low phosphorus iron in order to produce what was known as 'acid steel' or whether, increasingly typically, they used high phosphorus iron to manufacture basic steel. The steels that emerged were also classified into hard, medium and soft types, depending on whether they had high, medium or low carbon content. Then came the final stages, the making of semi-finished or finished products. The molten steel was normally cast into ingots and subsequently went through processes of rolling and forging in order to produce more or less finished products: for example, plates, rods, bars, rails, axles, tyres, billets and sections, and sheet and strip steel. Most of these formed, in turn, inputs for the principal user industries: shipbuilding, armaments, motor vehicles, tubes, and mechanical and electrical engineering.

It was this vertical form which imposed (and, of course, still imposes) the strongest pattern on the industry. It created a tendency for firms to develop vertically, by acquiring both supplies and outlets, as well as horizontally. But the choice on precisely *how* to expand was complicated by the way in which economies of scale operated. Technological change produced a constant impetus to growth both at any given stage and vertically. In general, there was a tendency for the most economic scale of plant at each successive stage to increase throughout the whole period. At the same time, this economic scale varied as between one stage and another, and the *rate* at which it increased also varied. In particular, the most rapid increases in the minimum efficient scale of plant tended to concentrate on the blast furnaces which made the iron, and on the steel-making furnaces. Theoretically, the most efficient scale of new plant should have been progressively introduced in association with adjustments of operating scales at the other stages in order to keep a balance. In practice, however, there were many obstacles to a smooth adjustment process, mainly because of bottlenecks over supplies (from earlier stages) and outlets (in terms of later stages), and because the question of what linkages of ownership, location and transportation were desirable – let alone available – between the various stages was highly complex. Difficulties arose over the relative cost advantages of links between (a) sources of coal and coke, and iron making; (b) iron ore supplies and iron making; (c) iron making and steel making, (d) steel making and the manufacture of semi-finished and finished products; and (e) the latter and the final user markets. The answers to these questions depended, of course, on the relative importance of each stage in the 'mix' of final costs: the location of available raw materials and markets, transport costs, and technological factors, for example the strong desirability of at least combining iron and steel making because of the greater ease of transferring hot liquid metal from blast furnaces to steel furnaces.

The industry's technology made it capital-intensive: labour costs were substantially outweighed by the costs of raw material sources, of plant and equipment for mining, processing and manufacturing, and of bulk handling and distribution facilities. Economies of scale at successive stages of the operation, also in terms of vertical linkages, made the industry relatively concentrated and oligopolistic. In most of the important markets a few relatively large firms controlled a high proportion of output and sales. Oligopoly, in turn, encouraged a quest for cartelisation, with firms seeking to band together to control output, markets and prices, by mutual agreement. But for various reasons such cartels had only a limited success. As theory would predict, they tended to work well only where the concentration was relatively great, the products relatively homogeneous, the consequences of price-cutting particularly ruinous and self-defeating for the firms, and the interdependence of competitors, therefore, very apparent. For reasons connected with foreign competition and UK trade policy, successful cartels also tended to require inter-firm co-operation on an international scale.

Thus technology and economies of scale laid a strong imprint on the iron and steel industry, endowing it with the key characteristics of verticality, capital intensity and high concentration. These factors, in turn, combined with others to produce a second overarching feature: the industry's marked sensitivity to economic fluctuations. Given the derived character of the demand for iron and steel, the influence of booms and slumps on the industry tended to be delayed. The industries using its materials were capital-intensive and particularly cycle-prone. The industry's own capital intensity meant that unit costs were especially sensitive to variations in the level of capacity utilisation. Hence unit costs fell dramatically in a boom just as they rose dramatically in a slump. Add to this the effects of the fluctuations on prices and the net result is clear: particularly sharp effects on current profits.

But this was not all. The industry's investment in new plant and equipment was also particularly strongly affected by the economic cycle. This was not only because of the profit variability just mentioned. The capital-intensity factor meant that the working of the accelerator principle, whereby changes in final demand induce greater effects on investment, was especially pronounced. Capital investment was lumpy, concentrated and particularly responsive to changes in economic expectations as a result of booms and recessions. During the period covered by this book these extreme sensitivities of both profits and investment were manifested in the postwar boom of 1918–20, the recession of 1921–3, the mild recovery of 1923–4, the worsening conditions from 1924 onwards which reached their climax in 1930–1, and the general recovery from 1933 to 1939, including a mild recession in 1937–8.

These characteristics applied particularly strikingly to the core activities of iron and steel making. It should be added, however, that in many related trades, closer to the final outlets, conditions were somewhat different. These outlying trades were technologically simpler in the main, generally less capital-intensive and oligopolistic, and often more profitable. They were important partly as a field for forward integration by heavy iron and steel firms, partly as breeding grounds for the contrariwise upstream movement of some enterprising and profitable firms into iron and steel making. Thus the constructional trade, where steel was used for bridges, dams and other major buildings, became an important activity for one of the three firms in our sample, Dorman Long. As a centre for semi-finished and finished steel products, Sheffield produced a business – Steel, Peech and Tozer – which became the nucleus of the second firm in our sample, the United Steel Companies. And it was another outlying sector, the steel tube industry, which fathered our third sample firm, Stewarts and Lloyds.

This last industry deserves further mention because of its continuing importance to Stewarts and Lloyds throughout the period. The steel tube industry manufactured tubes mainly for conveying gas, water, steam and oil, sometimes for structural purposes. Occasionally, still, it used iron rather than steel. Basically, it employed three main processes. Butt-welding, dating from the early nineteenth century, was usually employed for the tubes of smaller diameter. Lap-welding, a process originating in the 1840s, was applicable to larger sizes of tubes (or pipes, if that nomenclature is preferred – the distinction was not always clear). A third process, the manufacture of weldless tubes, was a late-nineteenth-century innovation whose products competed particularly with lap-welded tube, partly because they had better protective qualities for certain underground purposes. Although relative costs were slow to favour weldless tubes, the latter were destined to grow in importance. Compared with iron and steel, the steel tube industry was relatively atomistic and easy to enter, with low start-up costs and poor patent protection. The signs are that it tended to be more profitable. Its main centres were in the Midlands and on the Clyde.

## Historical and political factors

The factors so far mentioned affected both the iron and steel industry and its outlying sectors in a general way, regardless of period and country, and, indeed, in the main continue to affect it. To them must be added certain features which were distinctive to the industry in Britain up to 1939. This brings us to the third basic characteristic of the industry which needs to be strongly borne in mind: its long-term relative decline.

By the end of the nineteenth century the fulcrum of the industry, the combined manufacture of iron and heavy steel, was concentrated largely in three centres, Clydeside, the north-east coast, and South Wales. This reflected a number of dependences: on local coal supplies, on local iron ore supplies in the case of the north-east and South Wales, on sea-borne imports of iron ore, on the ease of exporting from nearby ports, and on accessibility to local heavy industries which were important users of iron and steel. Upstream from iron and steel making there were concentrations of iron ore mining and iron making in West Cumberland and parts of the Midlands, following the dictates of geology, and specialised iron firms persisted. Downstream the concentrations largely followed the more variable exigencies of markets: for example, re-rollers in the Midlands, manufacturers of steel railway products in West Cumberland and the north-east, tinplate manufacturers in South Wales, manufacturers of special steels in Sheffield (although these also benefited from good supplies of local scrap). This pattern of location and specialisation was built on largely mid-Victorian foundations. It reflected a preponderance of the steel-making processes which used acid or hematite iron ores; linkages with the older basic industries; a historic dependence on free trade; and sustainable but, on the whole, far from up-to-date scales of operation.

Over a long period leading up to 1914 these foundations had been seriously eroded. First, acid steelmaking was increasingly yielding to the basic process, introduced in the early 1870s, which was able to use iron ores with a higher phosphorus content. But whereas Belgium, France and Germany exploited good supplies of these, which were generally close to their steelmaking centres, Britain's appropriate iron ores were in the East Midlands, far from her existing steel-making centres, or on the north-east coast, distant from the industries best suited for the more ductile qualities of basic steel. Second, the new methods pointed to much larger scales of operation, particularly in coke manufacture, blast furnaces and rolling mills, and to increased vertical integration between processes. But these changes were generally pursued at a faster rate overseas for a complex mixture of reasons: their initially greater geographical concentration of production, their quicker economic growth and relative technical adaptiveness, a larger home base market in the case of the USA, tariff advantages and other factors. Third, a new pattern of industry demand for steel was slowly emerging, notably a relative decline in shipbuilding and railway building, and an increasing importance for engineering, special steels, and steels for new sectors like motor vehicle manufacture. For this new pattern of demand British production was ill-prepared and ill-located. Fourth, Britain's world trade position was worsening as a consequence of the high tariffs, cartels and economic nationalism of her main competitors. Her dependence on Empire outlets, which helped to secure some genuine respite in the 1905–14 period, only partially masked a long-term relative decline.

What could be done? For several reasons fundamental solutions were beyond the capacities of individual firms. For example, one important reason for decline, the overall shortage of industrial and technical training, demanded state action. So did much-needed improvements in the outdated and inadequate infrastructure of docks, waterways and railways, with which the industry was saddled. More controversially, the industry's problems highlighted certain dilemmas of political economy. Could the industry revive, let alone reform itself, without tariff protection? If tariff protection were granted, would the industry itself be capable of modernising, merging, shutting down inefficient plant and co-operating to secure external economies of scale on things like research, economic information and negotiation with foreign cartels? If state action were needed to secure these economic reforms, what form should such action take: persuasion, subsidies or legislation?

These questions, in turn, raise others no less complex. In so far as the economic reforms required to make the industry more internationally competitive succeeded, the industry would become more concentrated and cartelised. How, then, could the risks of price exploitation of the consumer be avoided? Would further public supervision be needed in order to prevent this? At the same time, would not purely economic reforms entail substantial social costs as a result of re-location and increased unemployment, particularly in isolated and dependent communities and during times of recession? How far should these social costs be taken into account and who should shoulder them? Together, these issues gave the industry its fourth key characteristic, that of being politically controversial.

Up to 1914 specific government policies for the iron and steel industry hardly existed. Public opinion was notoriously divided on the issue of free trade versus protection. No urgency was perceived for government measures to assist the reorganisation of the industry so as to meet increased foreign competition, and the idea of encouraging cartels, as a counterweight to the well-organised foreigners, would have been anathema in an atmosphere still dominated by Manchester School ideology. Consequently, the related idea of public supervision of an industry like iron and steel, to ensure that tariff protection and/or concentration were not exploited, hardly arose. Even less politically controversial propositions, to the effect that the industry should reorganise itself in some way, received only scant recognition. This lack of governmental and public interest partly reflected the fact that the industry itself was still divided on the tariff issue.

By the postwar period it was widely agreed in establishment circles that technical change had been too slow, particularly in the adoption of the basic process of steel making, that managerial efficiency and marketing efforts needed to be improved, that the industry was insufficiently concentrated to attain important production economies of scale, and that industry-wide co-operation was required for purposes of both central services and bargaining with foreign cartels. A broad diagnostic consensus on the industry's main problems was emerging. But a considerable number of years still elapsed before a public policy to assist these various purposes, let alone to deal with their various implications, emerged. Although the industry itself increasingly supported tariffs, the question of whether it should receive tariff protection remained entangled with the general political controversy on this issue; whilst on the equally politically explosive issue of how reorganisation was to be secured, with or without tariffs, opinion both inside and outside the industry continued to be divided and confused. By 1928 the Bank of England was beginning to be involved in moves toward rationalisation. But public policies only started to develop in a substantial sense after 1932.

It was in the 1930s, in fact, after more than a decade of depression and several decades of controversy over tariffs and industrial organisation, that the inter-related issues of political economy affecting the iron and steel industry finally came to a head. Effective tariff protection was introduced at last. Public efforts were made to get the industry to reorganise itself as a *quid pro quo* for tariffs, although with only partial success. Later in the decade efforts were also made to supervise pricing and investment policies. And government policies and public opinion increasingly exhorted sensitivity on a more perennial front, that of employment protection in depressed areas. These policies frequently conflicted, of course; and the means of securing them, representing a 'middle way' between the extremes of *laissez-faire* and coercion, raised important issues about the social control of business. Iron and steel became a test case for the idea of decentralised social co-ordination of private enterprise.

## The applicability of the theory

Before relating the characteristics just discussed to the idea of growth, efficiency and social action trade-offs and phases, it is first necessary to distance ourselves from certain debates among economic historians. Their controversies have centred around the hypothesis of a marked pre-1914 'entrepreneurial failure' in iron and steel, linked with wider theories about Britain's economic decline; counter assertions to the effect that the industry's performance was not really so bad, at least as understood within a neo-classical model; and discussions as to alternative explanations of decline drawn from a familiar list (the penalties of pioneer status, the collective character of the necessary adjustments, failures in capital markets, government policies, education or general social attitudes, etc.). Sometimes the influence of management is expressly minimised in these debates. Where it is highlighted, the emphasis has been on sweeping generalisations, showing a broad didactic swing 'from damnation to redemption', forcing the industry's entrepreneurs into stereotypes of rationality or inefficiency, progressiveness or complacency. But even apart from their pre-1914 emphasis, these controversies are largely tangential to this book because they deflect attention from the need to differentiate management.

Let us first consider the nature of the growth versus efficiency conflicts in iron and steel, and the problems they posed for management. To begin with it is necessary to take into account certain background considerations, first of all with regard to growth.

In the last chapter a distinction was made between growth by means of (1) increasing utilisation of existing capacity (the economist's 'short run'); (2) scaling up capacity in parallel with other firms in a general investment boom; and (3) pushing ahead of others in terms of relative size and competitive strength. It is with (3) that we are chiefly concerned. However, whereas in a rapidly growing industry a growth-hungry firm can achieve substantial expansion in its size not only absolutely but also relatively to other firms without noticeably displacing them, in the British iron and steel industry during our period this was not so. There was less chance of expanding boldly without treading on other people's toes. Of course, if recessions had the effect of shaking out the less capable firms, this would provide more elbow room. But on the contrary, marginal firms showed a strong capacity both to survive depressions and to exploit booms. This they did partly because their existing plant, however technically outdated, could still be physically adequate, partly because they were satisfied with low returns during recessions, and partly because of strong family, local and psychological pressures to survive. It was all the more difficult, then, for growth-hungry firms to achieve their aims without in some way tangibly affecting the position of other interests.

This factor, together with the industry's vertical relatedness, influenced the ways in which growth was pursued. It put a premium on the pursuit of acquisitions and mergers as a speedier and smoother way of dealing with rivals. Another consequence was a greater likelihood of growth through vertical integration, moving backwards to raw materials or forwards towards final outlets. If technological factors strongly supported such outward movement, so did the sheer unlikelihood of rapid growth within the industry's congested central nucleus. Yet another consequence was a greater pressure on the growth-hungry firm to expand overseas as a compensation for domestic frustration. These tendencies had important implications for the growth/efficiency trade-off, as we shall see.

The industry's economic sensitivity made ordinary investment decisions particularly difficult. Theoretically ideal solutions, of which firms were probably more aware than is sometimes conceded, suggested massive investment in new plant and equipment with the new-frontier, Penrose-type aim of combining higher efficiency and growth in the long run. Massive investment, though, did not merely involve the common-or-garden sacrifices related to financial conservation, curtailed dividends or resort to external finance. It also entailed the more than usually acute risk that a slump might all too quickly (a) cut short funds and delay completion; (b) turn the new plant, if completed, into an albatross because of its crushingly high unit costs at low output levels; and (c) compel reliance on existing outdated plant, which had meantime been neglected or even, perhaps, scrapped. On the other hand, again in theory, existing plant and equipment could be moderately improved with a stream of relatively minor investment projects. But in practice these came up against a host of interconnections, inflexibilities and indivisibilities, partly internal, partly infrastructural or external, which all too often forced one back to the *status quo*. And that stance was equally fraught with long-term risks, this time of consolidating inefficient operations and locations, wasting money on patch-and-mend expedients, and falling further behind foreign competitors.

It may be worth citing two management sources from our sample firms. In 1919 there was an exhaustive discussion at top levels in Stewarts and Lloyds about its Clydesdale iron and steel works. It was argued that this 'largely out-of-date plant' (a) could not operate profitably in competition with progressive competitors as it stood; (b) could not be disposed of because its output was still needed; and (c) could not be developed into a large-scale modern works because high capital costs could make it more expensive to operate than the *status quo*. This was a clear managerial distillation of many of the problems just described. The second source relates to the more serious difficulties in areas like Middlesbrough, difficulties which made even piecemeal improvements seem daunting. It is taken from an address by Francis Samuelson, a Dorman Long director.

> Here, in our old plants, our troubles begin. If we increase our engine power, we have not enough store power; we may add new stores if we have room for them, which often we have not; we may raise the height of the old ones, if they are strong enough for the new pressure. If we surmount the store difficulty, we find our mains and connections are not large enough to take the increased volume of air. If, by partial scrapping, we get over all these difficulties, we find that our yard is not equal to the increase in traffic. So that there really seems no half-way house between letting moderately well alone and complete scrapping; we can get a certain distance on the road to improvement, but not as far as we would like.[2]

Of course, it was still possible to improve efficiency in many other ways: by better co-ordination; by concentration of dispersed activities within a particular plant; by morale-boosting and welfare schemes with an eye to labour productivity; by good information systems on costs, profits, prices and market trends, designed to improve decision-making generally; by bulk buying of supplies; by pooled research facilities; by improved controls on stocks, working capital and investment; by centralised accounting and sales; and, not least, by improvements in management, ranging from the size and organisation of the board through decision-making processes to recruitment, promotion and development. Cumulatively, these policies could greatly help growth as well as efficiency. The trouble was, they required sacrifices: managerial effort, argument and conflict inside the firm, professionalism rather than amateurism, a move towards merit and away from nepotism in an industry widely affected by dynastic influences some of which were probably inefficient. Among the separatisms to be confronted were local and regional loyalties (even national ones in the case of an English–Scottish firm), identifications with old corporate images even where firms had merged, devotions to particular works, products or processes which might reflect, *inter alia*, simple technical conservatism. Two further obstacles were perhaps particularly strong in iron and steel: a semi-independent potentate position occupied by the works manager, and the resistance of vertically separate sub-units to the centralisation that was needed in order to exploit the economies of size and vertical organisation.

It may be argued that efficiency seeking could have been delegated downwards as a subordinate function, as 'operational' or 'executive' rather than 'strategic'. But this runs counter to the whole case for regarding as both central and persistent the problems of structure and control in large, diversified firms (see Chapter 1). The implication of that argument is that only top management could carry through tough rationalisation exercises, allocate production activities and investment funds, over-ride tough-grained separatisms and enforce central disciplines on sensitive issues like inter-unit supplies and transfer prices in a vertically related firm. Arguably, too, only top management could keep merit considerations ahead of purely family ones, when the two conflicted, or build up morale in an organisation of many thousands.

The time has come to make explicit the specific contribution of all these factors to the growth–efficiency trade-offs. In general terms their effect can be stated quite simply: it was to increase the conflicts between them and therefore also the likelihood of managerial polarisations towards one or the other or, in certain situations, neither. As foreshadowed in the last chapter, this drama can be hypothesised as occurring on three levels. First, on central issues where growth and efficiency were both sought, management were repeatedly compelled, usually unwillingly, to choose between them because of economic risks and time-lags. Second, the pursuit of one diverted managerial time and energy from the other. Third, each demanded a distinctive set of motives and abilities, and the chances of these distinctive sets combining within a particular managerial régime were, perhaps, small.

The first level was particularly confronted in investment decisions. The choice on whether to invest in a big way, at the risk of coming a cropper in a forthcoming slump, or to invest moderately, at the risk of further weakening long-term competitiveness, hinged on the view taken of the economic prospects. At best this was a matter of risk and probability, more likely a response to uncertainty. In practice, management would be swayed by the extent of its established inclinations towards growth. If a boom followed a major investment, growth and efficiency would eventually be harmonised at a higher level; if not, the firm would be left with the growth but not the efficiency and even, perhaps, if things were really bad, with neither. With a moderate investment policy, a limited type of efficiency would be served in an early recession just as a limited type of growth would emerge in an early boom; but long-run relative decline would be likely, overall.

Another time-lag, risk-related type of growth–efficiency conflict had to do with growth through merger. Although probably motivated in most cases by considerations of both growth and efficiency, takeovers and mergers tended initially to bring the former but not the latter. At any point of time suitable candidates would be few, so some elements of unsuitability about the match were likely. The merged units' overall size and some of the linkages created might be superfluous, undesired and potentially troublesome. The merger would probably involve major concessions to one's new partners, for example seats on the board, jobs for family members, security for long-serving officials, respect for established traditions. Much of this might have an adverse effect on efficiency. Not least, the widely favoured policy of backward mergers into raw materials posed distinctive problems. Especially in coal the available enterprises were likely to be small and economically marginal. As the best seams ran out a frequent recourse to more difficult workings and poorer qualities imposed lethal effects on costs. And the labour relations difficulties in coal were often acute, in striking comparison with their relative placidity, in the main, in iron and steel. In so far as growth meant mergers, therefore, more so than in expanding industries, it also brought in its wake substantial inefficiencies which would take time to resolve, even if the will and capacity to do so existed.

At the second level, time and energy conflicts were unavoidable, reflecting the enormous demands which growth and efficiency pursuits imposed. These conflicts could be exacerbated by failures in delegation and by the economic pressures to which iron and steel was prone, contributing to managerial fatigue, illness and even breakdown. They were highlighted by the demands and delays of travel: in complex iron and steel companies there was usually a four-way stretch between headquarters, scattered operating units at home, London as the industry's national centre, and overseas.

At the third level identified, the possibility of extensive managerial specialisation as between growth and efficiency can only be pursued through close observation of the individual firms in the following chapters. Suffice it to say here that the iron and steel industry poses the question raised in Chapter 1 in what is, perhaps, a particularly sharp form. Growthmanship required not just the usual panoply of ideas and appetites. Displacement effects, merger priorities and the internal politics of a concentrated industry meant that strengths in public relations, City contacts, lobbying, negotiating and industrial *haute politique* were also needed. But the pursuit of efficiency too was a major absorber of managerial enthusiasms and abilities. Within iron and steel it implied, on top of the typical underlying drives and values, such things as an intellectual grasp of complex and vertically diversified operations, from raw materials to final markets, and a ruthless streak to assist in tackling conservative, separatist and dynastic factors. The question of bias is inescapable. For it seems doubtful whether the two sets of attributes, needed for growth and efficiency respectively, could usually co-exist vigorously within a single dominant individual or even a small top group.

Let us conclude by referring briefly to the extra twist of complexity added by social action. As suggested above, pursuits of socially approved activities going beyond a firm's own growth–efficiency interests had particular significance in an industry which was strongly affected by externalities, social interdependences and, increasingly, public policies. Management had to decide how much time to devote to, say, national federation affairs in London or local public activities, and how much money to devote to charity (usually in small trickles). It faced the problem of how far to identify with the industry's collective interests on such matters as research or, by the 1930s, pricing restraint as a *quid pro quo* for tariff protection. By then policies to co-ordinate pricing and investment at national federation level and to respond to public and governmental influences in deciding these policies raised sensitive issues. How far could a firm go along with its colleagues and/or public policies on issues central to its private objectives? Should it be prepared to sacrifice some special corporate interests for the sake of political–economic advantages which were a risk and were to be shared proportionately or equally with others? What importance should be assigned to considerations of loyalty to the industry, public repute, safeguards against nationalisation, social duty? Seldom had large firms faced such complex challenges, at least in peacetime. As we shall see, the managerial responses varied strikingly from co-operativeness through reluctant acquiescence to separatism.

There were other issues on which social or political considerations often conflicted with sectional pursuits of growth and efficiency. For example, the question of environmental protection (the avoidance of river pollution, the restoration of land ruined by open iron-ore working); the introduction of joint consultation (where any productivity pay-offs might be debatable); the issue of what notice and compensation, if any, should be given to long-standing employees made redundant during the depression; the cases where duties to workers, dependant communities and public opinion seemed to demand the maintenance of sub-economic operating units. Although these issues of social action are less to the fore during the early part of our period, their high degree of definition at least by the late 1920s and the 1930s demands special consideration later on. Accordingly, Chapter 8 deals with a number of labour and local community problems, and Chapter 9 is devoted to relationships with government and public policies. Such issues are perennial and have caused wider debate in recent years, making the pre-1939 British iron and steel industry something of a pilot. Not the least of its advantages for our purposes is that it sharply demonstrates such perennial trade-offs between social action and the other pursuits.

## Notes

1   The most detailed general history is J. C. Carr and W. Taplin, *A History of the British Steel Industry* (1962). Also useful is T. H. Burnham and G. O. Hoskins, *Iron and Steel in Britain, 1870–1930* (1961). D. L. Burn's *The Economic History of Steelmaking, 1867–1939* (1940), still essential, is astringent and stimulating. P. W. S. Andrews' and Elizabeth Brunner's invaluable *Capital Development in Steel* (1951) serves as a partial antidote to Burn. Economic historians' controversies about the industry mainly before 1914 are pursued notably in D. H. Aldcroft, 'The entrepreneur and the British Economy, 1870–1914', *Economic History Review* (1964); D. S. Landes, *The Unbound Prometheus* (1969); and D. N. McCloskey, *Economic Maturity and Entrepreneurial Decline, British Iron and Steel, 1870–1913* (1973). A valuable and more recent survey, concentrating on the inter-war period, is S. Tolliday, 'Industry, finance and the state', Cambridge Ph.D. 1979. For a full list of published sources, including material on particular issues and the related trades, see the appendix on Source Material, Part (B).
2   S/GPC, 14.5.19, 13.6.19, 24.7.19, 29.8.19, 23.10.19. Francis Samuelson, presidential address, *Journal of the Iron and Steel Industry*, Vol. CV, No. 1 (1922) pp. 36–7.

# 3  Exuberance and caution

The period 1914–20 brought massive war production, temporary government control, large profits and major expansion projects within iron and steel. Financially, the industry did well. Psychologically, a heady mixture of patriotism and expansionism largely prevailed. The main significance of this period for our story is that it strongly tested corporate propensities for growth. The three firms in our study responded in strikingly different ways to that test, and the reasons for these contrasts take us at once into some initial exploration of the paths mapped out in the last two chapters.

The phenomena calling for explanation are fairly clear. A large, middle-aged business, Dorman Long, recovers something of its youthful growth rate. A new business empire is constructed at break-neck speed, the United Steel Companies (USC). In the third case, Stewarts and Lloyds, there is a notable lack of rapid growth despite some build-up of useful strengths. If growthmanship largely characterises the first two stories, caution dominates the third. These sagas also reveal differences in the exercise of business judgement. Avoidable errors appear to loom largest in the formation of the USC, least in the policies of Stewarts and Lloyds, with Dorman Long in a middle position.

In seeking to explain these contrasts it is necessary to review briefly the pre-1914 backgrounds of the firms, their market situations and managements. This exercise serves to introduce the firms. How far does it also support the ideas already advanced about decision making processes, key objectives and stages of corporate development? The answer is, only a little; but that little provides the groundwork for what follows.

## Growthmanship 1: Dorman Long's second spring

Of the three firms, Dorman Long was the earliest to grow large and the one most entrenched from the outset in the heartlands of iron and steel making.[1] From that central core it had ventured outwards along both of the paths suggested by vertical rationalisation: backwards to raw materials supplies, forwards to finished products. Geographically, the firm had remained relatively highly concentrated, on Teesside. A formidable embodiment of late-Victorian and Edwardian enterprise, by 1914 it was already one of Britain's leading iron and steel producers.

The firm's early history is largely interchangeable with that of its main creator, Arthur Dorman (1848–1931). Born at Ashford, Kent, and educated at Christ's Hospital, Dorman became apprenticed at eighteen to the owner of a small iron plant at Stockton-on-Tees. Having started out with some £1,500 of capital, in 1876 he formed a partnership with Albert de Lande Long to manufacture iron bars and angles for shipbuilding. Soon the firm was operating several works and challenging Belgian and German dominance in the iron girder trade. Moving from iron to steel as its basic material, in the early 1880s it installed open-hearth steel furnaces, first using iron made from imported hematite ores, switching later to the basic process which drew on local Cleveland ores. Capacity doubled between 1889 and 1896. By 1901 Dorman Long's labour force of 3,000 was producing about 180,000 tons of finished materials a year. It had already moved into a leading position in the long-established Teesside iron and steel industry whose principal creators had been Henry Bolckow and Lowthian Bell.

During the early 1900s the balance swung increasingly towards vertical integration and growth through amalgamation. The rate of growth in iron and steel production tapered off: it was about 55 per cent between 1906 and 1913. But forward movements intensified into constructional engineering and bridgebuilding, rolling mills, wire works, rails, and sheet works. Backward verticalisation embraced iron ore mining and collieries. Much of this was secured through mergers. In particular, control of Bell Brothers, an illustrious and highly diversified local firm, larger than Dorman Long and headed by the distinguished, now aged Sir Lowthian Bell, was obtained by 1902. A takeover of another substantial local firm, the North Eastern Steel Company (NESCo), soon followed. Overseas activities included subsidiaries in Australia and South Africa, and constructional work in a variety of countries, which brought increasing prestige. By 1914 Dorman Long had overtaken or absorbed its predecessors to become the dominant, although not unchallenged iron and steel firm on Teesside. More, with some 20,000 employees and annual finished materials outputs of around 400,000 tons, it was clearly in the national top league of the industry.

Dorman Long's progress had hardly been smooth. In the early 1890s a spate of external adversities was grimly felt. In 1901–2 a hiatus at the top reflected the short-lived retirement of Arthur

Table 3.1   *Company profits, 1910–28 (£,000s)*

|      | Dorman Long | SPT | S. Fox | Froding-ham | WISC | Stewarts & Lloyds |
|------|------|------|------|------|------|------|
| 1910 | 121 | 120 | 36 | 85 | 81 | 233 |
| 1911 | 181 | 93 | 31 | 117 | 81 | 272 |
| 1912 | 223 | 80 | 31 | 129 | 97 | 307 |
| 1913 | 258 | 126 | 56 | 137 | 57 | 346 |
| 1914 | 238 | 66 | 45 | 94 | 156 | 333 |
| 1915 | 405 | 107 | 66 | 137 | 134 | 356 |
| 1916 | 407 | 396 | 303 | 163 | 420 | 359 |
| 1917 | 406 | 111 | 378 | 396 | 209 | 386 |
| 1918 | 416 | 709 | 326 | 164 | 431 | 397 |
|      |      |      | USC |      |      |      |
| 1919 | 512 | | 1116 | | | 403 |
| 1920 | 743 | | 892 | | | 600 |
| 1921 | 414 | | 630 | | | 802 |
| 1922 | 160 | | (506) | | | 592 |
| 1923 | 207 | | 606 | | | 549 |
| 1924 | 505 | | 691 | | | 579 |
| 1925 | 231 | | 217 | | | 366 |
| 1926 | (179) | | 53 | | | 262 |
| 1927 | 273 | | (63) | | | 616 |
| 1928 | (353) | | N/A | | | 582 |

*Sources*: Dorman Long. D/AGMs, annual reports and accounts. SPT, S. Fox, Frodingham Iron and Steel, and WISC in USC, Net profits from 1870 (006/10/1). USC, Andrews and Brunner, op. cit., pp. 131–2. Stewarts and Lloyds, S/CB.

Dorman on grounds of fatigue: 'I have been constantly at it now for 30 years.' Financing the expansion meant relying on fixed interest borrowing and, still more, on large profit retentions (29 per cent of profits were ploughed back between 1904 and 1913), a policy which called for considerable nerve when economic conditions turned sour. Particularly in the early 1900s heavy investment in new plant and equipment coincided unpleasantly with a period of constricted demand. With the exception of one year, dividends were non-existent or minimal between 1902 and 1910, averaging only 3 per cent, which led to repeated complaints from shareholders. Only by 1911–13, as the stream of investment projects and technical advances bore fruit in increased profits (see Table 3.1), did dividends reach a more liberal average of 7 per cent. The large-scale mergers proved a mixed blessing organisationally. They left a heritage of elderly, non-executive, politically-motivated appointments on the Dorman Long board, there was much separatism, and in particular both Bell Brothers and the NESCo remained largely unabsorbed.

In all of this Arthur Dorman was the key figure as both chairman and managing director. He was a tall, imposing man who appears to have enjoyed his work, rewarded himself well and found it hard to delegate. The limited evidence suggests a towering personal dominance of decision-making, much resilience through successive economic cycles, and great courage in times of adversity. Dorman's speeches from the 1890s up to 1914 reveal a keen sense of international competition beside which Cleveland manufacturers might be 'a little too self-satisfied and prosperous' and to confront which he urged visits to overseas works to study 'the superior mechanical contrivances of other countries' and a consistently high rate of investment. Also emphasised were good marketing ('the very great importance of being in direct communication with the buyer') and diversification ('By widening and extending your manufactures and markets ... you are less liable to fluctuations').[2]

Dorman's overarching theme was expansion. Behind this lay an obvious zest for risk-taking, an enjoyment of substantial wealth, an enthusiasm for technical improvement: also, more or less in the background, were ideals of nationalism and empire, and a quasi-patriarchal concern to sustain Middlesbrough jobs and local pride. A cause so richly motivated and consistently pursued was bound to exact some price. Excessive optimism at times was an obvious result. On at least two occasions, in 1904 and 1909, Arthur Dorman freely admitted that he had been 'too sanguine' and 'mistaken' in forecasting good times immediately ahead. More subtly, expansionism might overshadow other concerns of importance to the firm. Dorman's defences against the occasional shareholder's imputations of administrative inefficiency were less sure-footed. The pursuit of economies of scale from the various mergers hardly featured in his statements. Nor did financial policy or the reinvigoration of top management.[3]

The period 1914–20 faced this large, long-established company, in some ways progressive, in others conservative, with formidable tests. The immediate exigencies of war brought a renewed phase of rapid growth. In its early stages Dorman Long produced about one-half of the high explosive shells used by the army. Arthur Dorman threw himself into the munitions drive with an energy and resourcefulness remarkable for a man now well into his sixties. For these services he received lavish rewards from the firm's coffers and, eventually, a baronetcy. The firm's profits increased substantially, although not phenomenally in real terms compared with others'. During the years 1910–14 profits had averaged £204,000. Between 1915 and 1920 they averaged £481,000, thus increasing nearly two and a half times in current money terms (see Table 3.1). However, from our point of view the central issue is how Dorman Long related these circumstances to a vision of the post-war economy, to plans and expectations particularly for the longer term. Here its response was consistent with long-established habits and preferences as well as with the generality of the trade. It opted robustly for a renewed phase of rapid growth.

Of Dorman Long's wartime expansion policies the largest was to prove the most two-edged. This was a plan for steel and plate making at Redcar, six miles from Middlesbrough. From early 1916 around £⅓ million was spent on land and plant at Redcar with a view to major developments to be assisted by the Ministry of Munitions. By early 1917 expenditure of about £4½ million on a complete new steel works, partially financed by a government loan, was envisaged. Although the scheme was cut down by the government, it remained extremely large. The new plant was to have 5,000–6,000 employees and a capacity of 300,000–400,000 tons per annum, an increase of at least 50 per cent on Dorman Long's pre-war capacity. By the end of 1917 a government loan of over £1 million was secured and well over £2 million had already been spent or authorised. Redcar cannot be called a major avoidable misjudgement comparable with others during this period (see the next section). On the information available in 1916–18 there were reasonable arguments for a massive new development, embodying the best practices of open-hearth plants on a large site further down the River Tees, closer to imported ores and better handling facilities for bulk materials, and free from the firm's existing largely constricted and misshapen sites in Middlesbrough. The main snag was to be the new plant's concentration on ship plates and the vulnerability of the underlying assumptions about shipbuilding industry prospects, although Dorman Long was far from being alone in this respect.[4]

The firm had long been anxious about its lack of self-sufficiency in both coal and iron ore. The Redcar project, combined with general optimism about the post-war economic prospects, now catapulted it into a scramble for increased raw material supplies. First, it spent some £300,000 on coal rights at *Betteshanger* in Kent, a development that was later to involve partnership with the Cowdray interests and much misjudgement (see Chapter 4). Second came the acquisition of an old-established Cleveland firm, *Sir B. Samuelson*, for £1.4 million in 1917. The main object, it seems, was to get control of Samuelson's ironstone mines in Cleveland, its eight blast furnaces (producing a quarter of a million tons of pig iron per annum), its collieries (contributing half a million tons of coal on average), and some two hundred coke ovens. Also, Samuelson's Newport Works was next door to Dorman Long's Britannia plant, offering some potential horizontal economies of scale. The firm had a respectable reputation technically, although it brought some duplication of Dorman Long's existing production and no access of new blood managerially. Recent profits, averaging about £200,000 in 1917–18, were a poor guide to peacetime prospects.[5]

The third step was the acquisition of the *Carlton Iron Company* for £950,000 in late 1919. This firm, which included an ironworks and a colliery in County Durham and a Scottish subsidiary, the Motherwell Iron and Steel Company, had experienced severe financial difficulties and, in its chairman's words, 'very unfortunate episodes' before the First World War, leading to a reconstruction in 1914. When purchasing Carlton, Dorman Long reckoned that a further £600,000 or so would have to be spent on developing the mines. The price paid for the firm was probably excessive, reflecting bullish expectations and a purchase at the crest of a boom. However, the saga of woes which Carlton was to bring could hardly be foreseen. In so far as they were perceived, its rough edges, like Samuelson's, probably seemed a necessary price to pay for an increased ability to exploit the firm's steel-making facilities (and during intermittent periods of high output subsequently, they did provide this).[6]

Mainly as a result of these developments, Dorman Long expanded more rapidly in 1916–20 than it had done through the prewar 1900s. In fact, it virtually recaptured a growth rate typical of its younger years in the 1880s. Expansion was helped by high profits but the main method of financing was by increases in capital. In spring 1918 there was an issue of £1.5 million 8 per cent preference shares related to the Samuelson takeover. In early 1920 an issue of £3 million ordinary shares, partly geared to the acquisition and development of the Carlton properties, increased the ordinary share capital from £4.5 million to £7.5 million. Dependence on loan or debenture finance was moderate.[7]

Dorman Long's economic optimism continued until late 1920. On top of the continued massive spending on the Redcar project in 1919 and 1920 (by late 1920 up to another £½ million because of delays and inflation) *plus* the £600,000 anticipated on the Carlton properties, 1920 brought further authorisations of capital expenditure: £106,000 on modernising the NESCo plant, and £230,000 on new steel sheet rolling mills at the Ayrton works. There was also a backlog of spending on Dorman Long's other works which it probably assumed could be safely left until 1921 or

1922. Optimism was reflected in higher percentage dividend distributions on ordinary shares than before the war, involving payouts of £204,000 in 1919 and £295,000 in 1920. It was reflected in the assumption that the £1.3 million government loan for Redcar would be repaid within three years of war victory and it was manifest in a characteristically exuberant speech by Sir Arthur Dorman at the April 1920 annual general meeting.[8]

In all of this, indeed, Dorman's dominance continued to be marked. Time and again the directors' minutes attest his hegemony. During the war it is Dorman who is left to decide a wide range of matters: the golden handshake for a retiring director; subscriptions to war funds; the conduct of negotiations with the Ministry of Munitions; the new housing scheme at Redcar; the terms of a large share issue. It is he who settles a subscription to the Cleveland Technical Institute, investments in overseas ventures, the purchase of scrap. In merger negotiations, it is true, Dorman is often seconded by Sir Hugh Bell, the deputy chairman, although in the Samuelson discussions in late 1917 only he is involved. Dorman's pre-eminence is also suggested by his position as both chairman and managing director; bonus shares of the profits way ahead of his immediate colleagues'; large shareholdings in the firm; his sheer time at the top. Indirectly, it is suggested by the probable lack of strong counterbalancing forces on the board. Sir Hugh Bell, Lowthian's son, a distinguished member since the turn-of-the-century merger with Bell Brothers, was a formidable figure (see particularly Chapter 9). But he was considerably occupied outside the firm and, it seems, was inclined to agree with Sir Arthur on the expansion strategies. As for the other directors, various combinations of absenteeism, old age, limited formal responsibilities and family relationships with either Dorman or Bell probably restricted their influence (see Chapter 4).[9]

The data suggest an absence of deliberation over the key decisions. Committees, it seems, were lacking. The decision to invest in various properties at Redcar, the start of a long and momentous chain of events, appears to have been a case of individual entrepreneurial initiative. Arthur Dorman and Sir Hugh Bell simply purchased £150,000 worth of works and mines on their own account. The Dorman Long board ratified the decision, took over the purchase on behalf of the firm and left the subsequent affairs to them. As this scheme escalated towards millions of pounds and top-level discussions in Whitehall, the chairman proceeded virtually on his own. The decision to acquire Samuelson for £1.4 million in 1917 appears to have been hurried as well as personal. The whole process up to formal approval of final

terms by the board took less than three months (although it
should be added that Samuelson's strengths and weaknesses
must have been fairly familiar). When the method of financing
the acquisition, an issue of £1.5 million new preferred shares, fell
through in March 1918, Dorman was authorised to pursue alternatives 'as may be thought best'.[10] How far this decision-making
mode can be termed purely *ad hoc* or 'incremental' will be
examined below.

As to its deeper provenance, Arthur Dorman's deep-seated
expansionism has already been referred to. Here was a product of
Victorian optimism, a founder-entrepreneur through his formative years, a temperamental risk-taker, and an enthusiast for
technical progress, job creation and the advancement of the
Empire. It is surely these longstanding characteristics rather
than the ephemeral politico-economic influences of 1914–20
which largely explain his growthmanship. True, government
money supported the Redcar scheme. But this project, critically
dependant as it was on the firm's own commercial forecasts and
plans for peacetime production, was emphatically not forced on
them. Indeed, Whitehall moderated it. True again, the embattled
chauvinism of the period drew on the nationalistic attitudes of
men like Dorman; but as one component of his expansionism such
attitudes were hardly fresh. It was not only a matter of temperament, conviction and ability. After all, Dorman was a veteran
warrior in the industry's central economic battles and it was
precisely this experience which seems to have taught him some
rational virtues in putting growth first. Repeatedly he had faced
economic adversity not so much by cutting or by pulling down the
hatches but rather by heavy capital spending. Tough though this
had been, it had brought great rewards in the ensuing booms and
a sense of enduring success overall. A habit of growth-seeking so
deep rooted and so long built up, through foul as well as fair
weather, was by now endemic.[11]

## Growthmanship 2: the USC's formation

As a major formation and a path-breaking amalgamation, the
creation of the USC was a more dramatic achievement than
Dorman Long's expansion during this period. It was grandiose,
verging on an apotheosis of growthmanship for the time. The new
USC represented an amalgamation about four times the size of its
chief initiating firm. Its scale immediately placed it in the forefront of the industry's top league. Its geographical sweep was
wide-ranging. From a compact nucleus in Sheffield the merger

took in other interests in South Yorkshire and made two great movements, southwards and north-westwards, capturing large bastions respectively in North Lincolnshire and West Cumberland. No less remarkable was the speed of the process. It was largely accomplished within two years, between 1916 and 1918.

Although the roots of this achievement lie deep, less is known about the antecedents of the various companies involved than is the case with Dorman Long's pre-1914 history. In particular, a key individual behind the process, Harry Steel, is relatively mysterious compared with Arthur Dorman. However, largely thanks to the pioneering work of Andrews and Brunner in 1950, much detail is available on the merger. Using more recently worked sources, too, a number of salient points can be pieced together about the decisions, the motives and the people.[12]

Within the context of this analysis several things seem reasonably clear. The creation of the USC involved a larger component of avoidable errors than Dorman Long's expansion. If the new combine's size and composition included great strengths, it also brought serious economic dangers and financial burdens, some of which at least should have been foreseen or avoided. The process of putting it together seems to have reflected a comparable measure of highly personalised decision-making. No less than Dorman Long's activities in 1914–20 it bears for good or ill the dominant imprint of a single individual. Moreover, there are signs of a highly charged managerial polarisation towards growth as the main propelling force. Of course, the motivational phenomena of the merger and the time included bullish economic expectations, nationalistic fervour, super-abundant cash from war profits and the rationale of vertical integration. But behind these lay more deep-seated managerial orientations: ones which have a bearing on our themes of trade-off, specialisation and corporate long-runs.

The main impetus came from a Sheffield firm, *Steel, Peech and Tozer* (SPT). This company, founded in 1875, had long concentrated on the production of high-quality acid steels. For this it used both pig iron imported from other areas and locally available scrap. The related activity of manufacturing semi-finished and finished products, particularly for the railways, was also important. In the 1900s semi-finished products made from basic steel were added. Profits averaged £134,000 between 1909 and 1914. Returns on capital were high and dividends generous. By 1914 the firm had a turnover of about £1.4 million and a total steel capacity of around 200,000 tons per annum, a respectable sort of medium size. By then it was largely controlled by second generation figures from the founding families: Albert Peech, J. E. Peech, William Tozer and, above all, Harry Steel.

Steel's formative experience appears to have been mainly in buying and selling. Within the USC his reputation largely remained that of a marketing man, a wheeler-dealer and negotiator. The few surviving photographs suggest a florid appearance in middle age, an Edwardian rather than a Victorian air, a hint of the *bon viveur*. Clearly, Steel was highly ambitious. Unfortunately, the deeper sources of this must remain largely speculative. Sheer temperamental ebullience? A zest for wealth and power? A marketing bias? The *arrivisme* of a relative newcomer in a city characterised by industrial aristocrats? A hunger for public recognition? Chauvinist attitudes? Or perhaps all of these? It is reasonably clear that Harry Steel was primarily a commercial man and not notably philanthropic. On the other hand, his ambition must have been multi-faceted and it undoubtedly included a public-political component. Steel played a leading part in the Sheffield industry's mobilisation for shell steel production in 1915. His references to national self-sufficiency as against dependence on 'the foreigner' for raw material supplies was probably no mere lip-service to a fashionable patriotic fervour. Intriguingly, Steel mentioned 'political influence' (whose, one wonders?) as an additional motive for large-scale mergers. His passionate advocacy of protection and his leading role in the steel industry's national counsels by 1918–20 (see Chapter 9) went well beyond purely corporate or commercial dictates.[13]

During the early part of the war SPT reached a point of major take-off. By 1916 the firm's expansionism and government sponsorship had combined to produce ambitious plans for new best-practice plant at Templeborough. These plans envisaged the mass production of basic steel on a scale more than double SPT's existing capacity plus big outlets for this production: a continuous mill for making standard billets and slabs. The latter were to be inputs for yet another development, a continuous strip and bar mill. Although partly dictated by wartime needs, these projects critically depended on anticipations of a long period of postwar prosperity apparently cherished by Harry Steel. Assuming that the Templeborough plant could be worked continuously at high capacity levels, production costs would clearly be low. Leaving aside for the moment how far the macro-assumptions were justified (see below), one immediate implication arose. The underlying economic expectations suggested that there would be serious difficulties with regard to the large quantities of coal and iron ore that would be needed. So did the immediate wartime frustrations over supply bottlenecks. Hence the idea of amalgamating with suitable raw material firms so as to control the supplies. It was from this chain of proximate circumstances that the rush to create a new combine largely emerged.

Of the substantial mergers which followed, three can be justified as reasonably sensible within the ambit of the overall plans, avoiding *ex post* criticism and taking account of the information available at the time.[14] These related to Samuel Fox, the Frodingham Iron and Steel Company, and the Rother Vale Collieries.

After an early period making wire for textiles and crinolines, and steel frames for umbrellas, *Samuel Fox and Company Ltd* had been an integrated steelmaker since the 1860s. Although continuing with specialities like wire, it manufactured billets and bars among semi-finished products, rails, tyres, railway springs and railway axles among finished ones; and it also owned a small colliery. Situated at Stocksbridge, a few miles from Sheffield, Fox was a smallish company, with an annual capacity of over 125,000 tons in 1917. It was owned and controlled by Francis Scott Smith, a friend of Harry Steel's, and, not least important, it was remarkably war-affluent (average profits 1909–14, £39,000; 1915–19, £268,000; an exceptionally large increase – see Table 3.1).

The *Frodingham Iron and Steel Company* originated as an iron producer in North Lincolnshire in the 1860s, exploiting the cheap iron ores of that area. It had manufactured steel since the 1890s, using the basic process, which was appropriate to the high phosphorus content of its local ore supplies. Owned by a local business family, Frodingham had been managed for many years by Maximilian Mannaberg, a technically brilliant and administratively capable executive of Czech origin. By 1917 Frodingham's steel capacity was about 180,000 tons per annum. Average profits were: 1909–14, £104,000; 1915–19, £215,000 (see Table 3.1). A possible flaw was the firm's involvement in an ambitious plan to develop steel making and plate making in the nearby Appleby Iron Company. This represented a heavy financial commitment. It, too, reflected economic optimism, including hopes for future shipbuilding outlets comparable with Dorman Long's over Redcar (see above).

The *Rother Vale Collieries*, situated between Sheffield and Rotherham, was producing nearly one million tons of coal per annum and developing modern coking plants which were to make it the largest coke maker in the Midlands. The firm's chairman and chief executive was Frederick Jones, originally a mining engineer, who had entered the company in 1879 during a financial crisis and who had largely built it up since.

There were considerable advantages in creating a union

between SPT and these three firms. Rother Vale Collieries' size, nearby location, efficiency and management were highly appropriate for the new combine. Samuel Fox's specialities, mainly in wire and wire products, offered some limited diversification of outlets for steel; its management was congenial to the SPT; its wartime profits provided further abundant cash for acquisitions. Fox's manufacture of billets and rails in the Sheffield area could prove a troublesome duplication of the SPT's unless rationalised between the two. It is possible, though, that the mere accretion of steel-making facilities was viewed as helping to justify the acquisition of raw material supplies which themselves needed to be large to be efficient. The acquisition of the Frodingham Iron and Steel Company provided a large stake in the cheap iron ores of North Lincolnshire, thereby safeguarding supplies of the basic ores needed by SPT and Fox. It also secured a first-class steel-manufacturing capacity on the same site and an outstanding chief executive. The failure to predict that the associated Appleby project would become a financial millstone can be criticised on the basis of rather rigorous criteria (see below) but this does not call in question the overarching validity of the Frodingham merger.

The new combine's fourth major component, the large interests it acquired in West Cumberland, are another matter. Whatever their eventual social advantages (to local employment and national defence), these operations were to prove an albatross commercially. The disadvantages of investing heavily in West Cumberland iron and steel making should have been clear in 1917–18. These disadvantages included deficiencies of the local coal and coke, high costs of pig iron made from the local hematite ores, and the unreasonableness of local holders of royalty rights. They reflected the retreat of acid-steel making on which the local pig iron depended, the vulnerability of the steel rail market (a principal outlet for acid steel), and the area's isolation and its lack of a deep-sea modern dock. Such difficulties had already caused one major firm, Cammell Laird, to withdraw from the area. Many of the adverse trends were clearly evident in statistical data available by 1917. Both the long-term trend towards basic steel making and the static character of West Cumberland's output were familiar factors. The *Workington Iron and Steel Company*, formed in 1909 to confront the problems, had had very limited success. It is true that the WISC possessed a high degree of vertical integration, some admired technical skills, little local competition and a substantial size (with some 5,000 employees and annual outputs of about a quarter of a million tons of hematite pig iron and about 160,000 tons of acid Bessemer steel). However, its net profits had been slight even before the war (see

Table 3.1). For many years its chairman, Sir John Randles, had voiced concern about the firm's vulnerability and by 1917 he was warning about 'poor prospects' unless it could join up with a bigger outside unit.[15]

The terms of the merger were generous to the WISC interests. All the WISC shareholders gained higher ranking stock in a bigger and potentially much stronger concern. Of a total capitalisation of the USC of about £10.2 million, some £2.6 million was accounted for by the WISC operations, certainly a high figure relative to previous peacetime earnings. No wonder Randles praised the deal as 'a very excellent one for the shareholders'.[16]

Why, then, did Harry Steel make this major strategic move? Although he mentioned benefits *vis-à-vis* technical knowledge and selling, his main motive seems to have been a desire to safeguard supplies of hematite iron ore for the Sheffield steel plants. This was commercially rational in so far as the USC's Yorkshire plants would continue to make acid steel products which required this material: during the war they had probably particularly suffered from undue dependence on foreign supplies. But according to Andrews and Brunner, SPT's and Samuel Fox's 1917 plans required up to about 40,000 tons of acid pig iron. This was a small demand relative to the WISC's capacity of up to a quarter of a million tons so that clearly the latter was expected to continue as a major manufacturer of steel products, using its own pig iron; and much of its sales would be in the acid steel specialities also produced by SPT and Fox.[17] It is hard to acquit these two firms (and more particularly Harry Steel, see below) of avoidable misjudgement here. Even apart from their general economic optimism, the specific market assumptions were ill-considered. The WISC merger would massively increase the new USC's involvement in the problem area of hematite iron ore and acid steel products, and it ignored the evidence of high costs, meagre prewar profits, management problems and local fears. A tentative diversification plan might have outweighed such omens, at least partially, but there is no sign of one. Even if the WISC purchase is conceded as necessary to secure steady supplies, assuming a continued heavy commitment to acid steel, the high price paid, let alone the heavy debenture obligations created, would surely have struck a detached contemporary observer as excessive.

Having assembled its four main units, the new USC proceeded in much the same robustly expansionist frame of mind through 1918–20. This mood was evident in an additional large investment in the Appleby project, the acquisition of still further interests in West Cumberland (designed to round out the position there), various modernisation projects and a continued optimism well into 1920, shown, for example, by 'heavy purchases of scrap in anticipation of a good market for steel for a considerable period ahead'. It was evident, too, in a continuance of high payouts of dividends until autumn 1920, two share issues (in 1918 and 1920) and, very important, heavy additional debenture borrowing.[18]

The decision-making behind all this appears to have been largely Harry Steel's. As chief executive of SPT, he was probably very much its dominating figure, also judging by the later careers of his immediate colleagues there (see Chapter 4). Probably, too, Steel was a stronger personality than his friend and first main ally in the mergers, Francis Scott Smith of Samuel Fox, whom the later records suggest was a doughty defender and administrator of his old firm but perhaps not much more. The board of the new USC, which included the formidable Frederick Jones of Rother Vale Collieries, did not meet until April 1918 and is unlikely to have become an effective body straightaway. The equally if not more formidable Maximilian Mannaberg of Frodingham Iron and Steel did not join the board until 1922. Frederick Jones's son, Walter Benton Jones, had not yet become a substantial force. So there appears no reason to dissent from the popular contemporary view, reinforced by Andrews' and Brunner's interviews in 1950, that Harry Steel was probably the dominant force both in the formation of the USC and during its first two years. It is also hard to resist an impression of rush and lack of deliberation in the decision-making.[19]

Both points are evident particularly in the decision to invest massively in West Cumberland. Harry Steel seems to have been the prime mover here. None of his principal colleagues was involved in the merger discussions, as reported in the WISC directors' minutes, or in the ensuing statements to the firm's shareholders and the press, in April 1918. The core process of investigation, valuation, negotiation and final agreement with the WISC directors was hurried through in scarcely more than three months. The financial advice of the senior accountant, Sir William Peat, hardly compensated for a lack of detailed technical, market and economic evaluation. There is no sign of consultancy advice or involvement on the part of the new USC's foremost steelmaking expert, Maximilian Mannaberg. It was only in June 1918, *after* the event, that Frederick Jones reported on 'a very thorough investigation of the ore properties of the WISC'.[20]

We revert to the question of underlying managerial thrusts and their biases as well as strengths. Harry Steel's growthmanship and its probably long-established character have already been referred to. It is doubtful, to say the least, whether this expansionism was matched by any comparable concern or aptitude for efficiency. Even if SPT had been efficiently organised (and evidence here is unfortunately lacking), this would hardly prepare someone for the task of running so large and complex a concern as the USC. Harry Steel's background is unlikely to have encouraged much involvement in corporate organisation, management development, information systems or financial controls. His explicit objective was for the new combine to be a loose confederation, a grouping of friendly but still basically separate units. That concept applied even to the problem-ridden WISC, as shown, for example, in a remarkable statement to the WISC shareholders in April 1918. On the brink of finally consummating his headlong courtship of upwards of one-quarter of his new empire, Steel gave no hint of cleansing or even investigating its weaknesses of which, as already suggested, he seems to have been largely unaware. On the contrary, he disclaimed any desire 'to try to interfere with the management and control of the company'. During Steel's two-and-a-half years as chairman and chief executive of the new USC there is virtually no sign of any start towards reorganisation or reaping the economies of scale. Nor did he have much time available for such pursuits. He seems to have been largely absorbed in still further acquisitions and expansion projects, and also in negotiations with Whitehall over the financing of the Appleby scheme. Social action, too, claimed much of his energy, with a fast-growing role as an industrial statesman and leading politician within the national federation. None of this is intended to diminish Harry Steel's achievements as an empire-builder but rather to emphasise the concentrations and sacrifices of other pursuits which appear to have been involved.[21]

## Stewarts and Lloyds' prudence

To move from the USC to Stewarts and Lloyds during this period reminds one of being on a guided tour of stunningly contrasted architecture. It is like moving from a massive baroque construction to a simpler neo-classical building. If the former is solid in its main parts but over-extended, its ornate façade masking some less impressive architecture behind, the latter is relatively sober, disciplined and spare. In fact, the 1914–20 development of Stewarts and Lloyds poses some particularly sharp contrasts with both the other firms. Whereas optimism and growthmanship dominate the Dorman, Long and USC sagas, albeit in different degrees, Stewarts and Lloyds' policies are largely steeped in caution. Expansion is moderate, forward plans are hedged, significant growth opportunities are refused. Where misjudgements are recognised after the event these are partially retrieved.

Again, such surface phenomena are easier to delineate than the underlying forces. Superficially, the most obvious explanation for the contrasts might seem to be the nature of Stewarts and Lloyds' industrial activities as compared with iron and steel. Certainly, any analysis must pay careful attention to the various characteristics of the steel tube industry introduced in Chapter 2. However, it is arguable that Stewarts and Lloyds' prudence mainly derived from a different type of management: a different configuration of corporate politics, power distributions and basic orientations of personal values and abilities.

As before, a brief survey of activities before 1914 is needed.[22] The firm of Stewarts and Lloyds was formed in 1903 as a merger between two older companies, both highly vigorous. *A. and J. Stewart and Menzies*, on Clydeside, was mainly descended from an 1861 formation by Andrew Stewart, a canny and ambitious businessman. Among other distinctions, Stewart subsequently endowed the Adam Smith Chair of Political Economy at Glasgow University. His firm built up its marketing, acquired a number of tube competitors in Scotland, diversified its product range. It also extended backwards into iron and steel making for the sake of its own supplies, acquiring the substantial but not overly efficient Clydesdale iron and steel works in 1890. The business of James Menzies and Company was absorbed in 1898. By 1903 the combined firm of A. and J. Stewart and Menzies had come to dominate the Scottish steel-tube trade and it also had a strong hold in export markets. Profits over the previous five years had averaged well over £100,000 on a capitalisation of slightly over £1 million.

*Lloyd and Lloyd* had been founded in Birmingham in 1859. The firm sprang from a ramified, originally Quaker dynasty whose other (partly complementary) interests spanned iron ore mining and manufacture, banking and politics. Another family, the Howards, also became involved. Lloyd and Lloyd expanded its works and offices inside Birmingham, and acquired the important Coombs Wood Works at Halesowen, nine miles away. By 1903 the firm was of comparable size to its Scottish counterpart. By contrast, however, it had concentrated on a narrower product range for a larger domestic market, and it had an outstanding record of technical progress. Thus the two firms' interests were partly complementary as well as conflicting. A longstanding rivalry between them came to a head in 1900 when Lloyd and Lloyd purchased a Clydeside business with the aim of manufacturing inside Stewart and Menzies' home territory. Negotiations to avoid an all-out war resulted in the 1903 merger.

The new Stewarts and Lloyds combine made steady, if unspectacular progress up to 1914, nearly doubling its output of finished tubes and fittings, also its production of steel in Scotland. Further acquisitions were marginal. Instead, growth came largely from fairly heavy capital expenditure and vigorous marketing. By 1914 Stewarts and Lloyds were exporting nearly one-half of their finished tubes and fittings and accounted for about one-half of the entire UK output of mainstream tube products. The firm's capitalisation was £1.8 million. Profits had been stable and high (see Table 3.1).

Further keynotes were financial abstinence and some concern with administrative efficiency. The former theme emerges across a wide spectrum: rising allocations to both depreciation and reserves, restraints on borrowing (held at about 18 per cent of capital employed), moderate dividends, and levels of directors' remuneration that were also soberly spare. The company's neoclassical, almost puritanical style is strongly evident here. It emerges, too, in some post-merger efficiency improvements. True, a substantial rationalisation of production facilities was to be long delayed (see Chapter 4). But the inter-plant allocation of orders was improved. Overheads were reduced. Export sales administration was concentrated in Glasgow and Birmingham. Accounting systems were tightened up. More subtly, the restraint over further mergers partly reflected, it seems, a preference for absorbing existing resources rather than taking on additional problems viewed as difficult to run efficiently.[23]

Why did Stewarts and Lloyds pursue this cautious mixture of moderate growth and partial economising between 1903 and 1914? More important for our purposes, why did broadly the same mixture persist through 1914–20? No doubt we have to look to the firm's position in the steel tube trade for part of the answer. A company with about one-half of national output already could not expand that easily. A company so much involved in exports and international competition and one so vulnerable to small, price-cutting competitors at home had to be cost-conscious. It needed to exercise some restraint over pricing and profit-taking, partly also for fear of new entrants to the trade. The same influences, of course, largely pushed Stewarts and Lloyds into its persistent search for cheaper supplies of steel, the biggest component of costs. But these market-environmental factors come nowhere near explaining the firm's extreme caution over such growth opportunities as occurred, its pessimistic streak, its deliberative style, its interest in organisational efficiency.

A crucial managerial factor was undoubtedly Stewarts and Lloyds' character as a relatively new merger between two firms from different traditions and cultural settings on opposite sides of the Border. In such a *mariage de convenance* the highest common factor of agreement tended to be limited. When Lloyd and Lloyd's chairman, Henry Howard, was reported on the objectives of the merger the tone was severely narrow: 'There had been one uniform result from all this fighting to the shareholders of both companies, that whether victorious or not it had diminished their profits. This competition would for the future cease to exist, and this concentration of practical experience should enable home and foreign competition to be more easily dealt with.'[24] At local levels some larger enthusiasms and visions of business purpose doubtless continued. So did national, civic and cultural influences among the Stewart échelons in Clydeside and those of the Lloyds and Howards in the Midlands. But any new corporate binding force among such disparate transnational elements was still likely to be somewhat thin. Common technical enthusiasms could not provide much of an emotional drive. For steel tubes lacked the production mystique and overtones of industrial aristocracy enjoyed by heavy iron and steel. Even by 1914–20 there are few signs of either the visionary concepts or the public stances characteristic of an Arthur Dorman or a Harry Steel. In such a context top management was unlikely to polarise on either growth or efficiency (let alone social action); it was more likely to follow a restricted consensus, a sober balancing act.

The situation would have been different if a single strong individual dominated the board, but this was not the case. The reputation of John G. Stewart (1862–1925), chairman since the 1903 merger, may have suffered from some posthumous denigration inside the firm. An 'in-house' history says nothing positive about him and implies some weaknesses: a limitation to the production side, a lack of judgement. It takes as 'a revealing picture of J. G. Stewart's personality' a paper he wrote in 1908, a rather extremist argument for large-scale backward integration into coal which was quickly shot down by his brother, T. C. Stewart, and not acted on. This seems unfair. In fact, a moderate defence of J. G. Stewart can be mounted (see Chapter 4). But even if the implied criticisms are unjust, it is probably true that his power was highly qualified and that Stewarts and Lloyds' leadership in 1914–20 approximated to a group of near-equals. Six men were important, judging by the minutes and other documents, their designation as joint managing directors in several cases, their membership of key committees and their remuneration: T. C. Stewart (general policy), A. W. Lloyd (commercial policy), G.

A. Mitchell (material supplies), Joseph Howard, Jr, J. H. Lloyd and, of course, J. G. Stewart (tube making). Together with Henry Howard (the elderly deputy chairman) and R. M. Wilson (another senior statesman), this group included a wide range of abilities and seems to have included significant cross-currents on matters of policy.[25]

From all this a deliberative style of decision-making tended to emerge. For example, decisions as to the best centres for iron and steel making were pondered over for years; in 1914 they were postponed because of lack of information; in 1916 the technical properties of the respective iron ores were independently tested by a distinguished metallurgist; while takeover negotiations with a particular firm took nearly two years. When the future of the firm's existing Clydesdale Steel works was once again thrown into the melting pot in 1919, an intense, sometimes heated debate ensued on how inefficient that works really was and on what should be done with it. The discussion embraced all the available alternatives: disposal, pure supplying for tubes or wider competition with Scottish steelmakers, and, in the last two cases, the extent of required modernisation. This debate, like some earlier ones, was marked by an exchange of detailed papers between the protagonists.[26]

Given all these characteristics, it is hardly surprising that Stewarts and Lloyds in 1914–20 presents a striking case study of circumspection, a microcosm of caution. Five main policy phases manifest this theme.

(1) *The initial selection of new supply sources.* Ideally, Stewarts and Lloyds needed to be able to draw on cheap iron ores, to operate an integrated iron and steel works near those ores at the highest level of efficiency, and to locate tube making in the same position so as to minimise bulk material transport costs. Although that optimal combination was not approximated until the 1930s, the firm was already inching towards it. Its existing Clydesdale steelworks, controlled since 1890, was inadequate: distant from suitable iron ore supplies and of sub-optimal size. For many years Stewarts and Lloyds had been quietly investigating possibilities of cheap basic iron ore from both Lincolnshire and Corby, Northamptonshire. By 1917 its attention was concentrated on two small companies, the North Lincolnshire Iron Company (NLIC) at Frodingham, and Lloyds' Ironstone Company (LIC) at Corby. Either or both of these could be purchased to supply pig iron for the Scottish and English tube plants, with the additional, more radical possibility of building a new, integrated iron and steel works.

Ironically, in view of its future critical relationship with Corby,

Stewarts and Lloyds decided, in spring 1917, in favour of the NLIC. An outside expert advised that a new steel works should use the open-hearth process, for which the lower phosphorus Lincolnshire pig iron would be more suitable, and a financial appraisal suggested that, whilst operating costs would be similar as between the two projects, the immediate outlay *vis-à-vis* the NLIC would be lower. The circumspection persisted. Stewarts and Lloyds limited its acquisition of NLIC holdings to 70 per cent of the shares. After long negotiations and many calculations it paid a price, £175,000, which was hardly a major commitment of its resources. And although the central idea was to erect a steel-works, the option of withdrawing from this was retained.[27]

(2)   *The flexible alteration of that decision.* In fact, further caution and difficulties over infrastructure services delayed the construction of the steelworks. By January 1920 Stewarts and Lloyds reckoned that the (inflation-boosted) capital cost would be £2.4 million and, pessimistically, it expected a steel trade recession by the time a new steelworks was on stream two or three years later. The NLIC's existing ironworks could not be disposed of and was to create much trouble. But withdrawal from the steelworks project was decisive and, as things turned out, prudent. There were still resources to pursue an alternative source of steel, the small firm of Alfred Hickman Ltd. Although much of this company's plant at Bilston was antiquated, it was felt to be capable of 'giving a fairly large output at a reasonable cost under present conditions' with scope for 'very considerable extensions and improvements'. Significantly, Alfred Hickman had meantime come to possess the LIC with its Corby ironstone, previously rejected by Stewarts and Lloyds. In view of all this the acquisition terms – an exchange of shares and a cash payment of £325,000 to Hickman's shareholders – were probably reasonable. Thus Stewarts and Lloyds not only avoided burning its fingers badly, but it was now also assured of steel supplies for its English works and had come to possess potentially valuable Northamptonshire iron ore resources.[28]

(3)   *The refusal of tube-making expansion opportunities.* Mean-time the chance of acquiring a formidable competitor, the British Mannesmann, was turned down in 1916 because the asking price of over £530,000 was felt to be too high. Acquisition possibilities *vis-à-vis* the Clydeside Tube Company and the Scottish Tube Company were refused in 1918 and 1920 respectively. There are glimpses, too, of a circumspect pricing policy on tubes. The aim was to avoid two main pitfalls: cut-throat competition (damaging in the longer term) and high prices (inviting public criticism of 'profiteering' and endangering the renewal of tariffs). But if rapid

advances in the tube market through takeovers or aggressive pricing were viewed as too risky, as was forward diversification. In July 1919 the chance of undertaking the manufacture of corrugated furnaces was turned down. There was little relish for fighting 'a powerful combination' which controlled that trade and which could, it was thought, retaliate by hurting Stewarts and Lloyds' existing tube business.[29]

(4) *Financial conservation.* From their high prewar base, profits rose moderately early in the war, then increased more decisively by 1918–21 (see Table 3.1). The firm's commercial caution and political fears involved a careful discretion over pricing, as already mentioned, and therefore over profit-taking also. But financial caution also probably led to profits being understated. They were probably higher, allowing for hidden reserves and ultra-careful hedging against war tax liabilities. Yet dividend rates were kept stable (only going up once in 1920 and then very moderately) so that in 1919–20 hardly more than one-fifth of net profits was paid out in dividends (as compared with, say, Dorman Long's two-fifths). Reserves were greatly increased. The capital structure was kept spare and conservative. Almost certainly, substantial long-term loan finance could have been obtained. Not only was this refused, but in 1918 increased profits were used to wipe out a venerable low interest debenture of £350,000, thus actually reducing the long-term debt ratio. Over-capitalisation was avoided like the plague. With issues of deferred shares in 1918 and 1920 issued share capital increased to £3.6 million, a far from untoward rise on the 1914 figure, given both the earnings trend and probable under-capitalisation before the war. Thus fidelity to the firm's classical financial puritanism was maintained.[30]

(5) *Economic pessimism.* Stewarts and Lloyds responded quickly to any bad economic signs and was well ahead in anticipating a recession. At the first glimpse of some impending economic troubles in early 1919 several investment projects were rejected and the forward buying of raw materials was curtailed. Forward planning included possible bad outcomes and contingency plans, although profits in 1920 turned out better than expected. In January 1920 fears of later recession strongly influenced withdrawal from the NLIC steelworks project, as we have seen. In June 1920 negotiations to purchase a colliery for £600,000 were called off, and in November 1920 there was further gloom about 'a danger zone immediately ahead' even though, again, profits in the following year showed a substantial increase. Once more the contrast with the other two firms was marked.[31]

**Final comments**

From this outline of the comparative 1914–20 experiences and their antecedents several points emerge: first, on decision-making processes. If none of the firms practised a precise optimising calculus neither did they fit the extreme 'behavioural' view. Their policies were far from 'incremental' in the sense of a mere *ad hoc* reactiveness to outside pressures or a preference for minimal changes and 'getting by'. All three managements had strategies which were fairly consistent, comprehensive and also radical in the sense of going well beyond mere survival. The 'behavioural' view faithfully (and, it may be said, tritely) echoes the obvious facts that methods of implementing the strategies changed, reflecting a fluid environment, and that decisions on investment projects and acquisitions had to be taken within severe time and information constraints and often under stress. But although each such decision, looked at separately, may seem to reflect more hunch than calculation, the longer period data tend to suggest at least some previous knowledge and planning, as for instance over Dorman Long's acquisition of Samuelson, and the USC's of Samuel Fox. A restriction to management's 'current' descriptions, let alone the precise recording of their time-specific decisions in minutes and elsewhere, understates both consistencies of purpose and implicit or even explicit calculus. As suggested in Chapter 1, a more useful contribution to the 'rationalistic versus behaviouralist' debate is to pinpoint comparative degrees of 'rationality' and 'error' within the zone between these two conceptual extremes.

At the most excusable end of this intermediate spectrum come Dorman Long's and the USC's anticipations of a postwar economic boom. Even here there is some difference of emphasis. Arthur Dorman's brand of 'bullishness', tempered as it was by long experience of cyclical stress, was probably less marked than Harry Steel's assumption of a 'long' period of prosperity and extravagant political forecasting ('Hitherto in this country the view of the Government has been that the industry could look after itself. That view, I think, has gone for ever').[32] The important point, however, is that these men's optimism about economic revival and diminished foreign competition after the war echoed a quite general contemporary view. The echoing can be criticised only on the ultra-rigorous and arguably unsustainable view that individual firms' macroeconomic forecasts should excel those of established élites generally, of government and the City as well as of industry. Fewer such allowances apply to another form of collective euphoria, the assumptions made about a particular market outlet like shipbuilding. In anticipating a specific boom in shipbuilding (in the context of a general boom), both Dorman Long's Redcar project and the USC's Appleby project merely conformed to a fairly general market optimism. The 'merely' is significant and hints at some astringency. After all, it should have been easier to find some contrary opinions from elsewhere; the decision-makers were closer to the specifics of such a neighbouring industry; and arguably a single sophisticated firm *could* have calculated more warily than the rest of the pack to which it belonged. Again, it is probable that Dorman Long was a shade more inclined to hedge its bets over Redcar than the USC was over Appleby but the evidence here is limited.

Next come those errors of commission which strongly reflected assumptions about sectors in which a firm was already heavily involved. Basically, these were mistakes about specific iron, steel and coal expansion projects or acquisitions. Here it is not good enough to call macroeconomic forecasting difficulties to the firms' defence. For important though the macro-assumptions were, estimates specific to their own industry or its immediately adjacent trades were also critically involved. But within this category different shades of rationality emerge, reflecting varying contours of previous experience, deliberation over the choices and proportions of corporate resources committed either immediately or in prospect. For example, Stewarts and Lloyds' errors over the LIC in 1917 appear less serious than those of Dorman Long over the Carlton Iron Company in 1919. The former firm had less direct experience of the sectors involved and did a lot of detailed research over a long period. Also, it spent less money straightaway as a proportion of its resources and hedged its bets more carefully on future commitments. Both cases, in turn, appear less serious than the USC's (Harry Steel's) misjudgements over West Cumberland. These latter, the available evidence suggests, seriously offended virtually all the criteria.

The final main category of errors, those of omission, is clearly the hardest to evaluate. In this field the economic concept of opportunity cost, the benefits of the best available alternative foregone, meets its severest operational tests. Also, serious historiographical problems arise because of inadequate evidence on the alternatives considered by a firm but rejected. The issue comes to a head over Stewarts and Lloyds' saga of caution and whether this was excessive. Theoretically, that firm could have pursued some combination of (a) increasing its share of domestic tube-making (for example by acquiring Mannesmann); (b) investing in facilities for the more innovative process of weldless tube-making already followed by German competitors; (c) effecting more ambitious iron and steel takeovers; whilst (d) drawing on its large reserves and/or increasing its modest borrowing ratio so as to achieve such ends.

All these courses had some cogent reasons stacked against them in the decision-makers' minds: (a) would have involved large expenditure at the cost of pursuing (b) and (c). It would have provided no reasonable assurance of a commensurate increase in market control, given a continued tail of small competitors and easy entry conditions. Then (b) incurred some respectable, if unadventurous objections suggested by J. G. Stewart in a detailed paper in 1918. A decisive move towards weldless tubes would be commercially justified only when they could be 'produced of as reliable material and at about the same cost as a welded tube', although 'that day ... is always growing nearer'.[33] As for (d), it would have constrained the firm financially in a forthcoming recession. Only on (c) does the charge of excessive caution appear strong. A technical misjudgement over iron ore suitabilities and, perhaps less excusably, some narrow criteria of immediate capital cost minimisation caused Stewarts and Lloyds to get cold feet over the Corby alternative (later to be the brightest jewel in its crown; see Chapter 6).

As for the reasons for Stewarts and Lloyds' consistent caution and the other two firms' consistent expansionism it is managerial factors which have been stressed. This is not to deny the role of changing economic and political circumstances between 1914 and 1920 nor the reality of the market–technology contrasts between Stewarts and Lloyds and the largely iron and steel firms. It is, however, essential to recognise the importance of underlying managerial dispositions, their convergence on certain broad patterns, and their amenability to systematic explanation.

In any such explanation the question of managerial power distributions must play some part: diffused in Stewarts and Lloyds (reflecting both the Scottish–English factor and an apparent plurality of strong individuals), concentrated in both Dorman Long and the formation of the USC (reflecting the personal hegemonies of Arthur Dorman and Harry Steel). The deliberative, dialectical style of decision-making in the former case is seen as an obstacle to extreme courses generally and bold expansionism in particular. The personalised decision-making in the other two cases is seen as an encouragement to managerial biases or specialisations.

As to the nature of the underlying managerial orientations, something of the rich motivational diversity and long-established character of growthmanship has been glimpsed in both Arthur Dorman and Harry Steel, particularly in the former. Hints have emerged that this growthmanship probably involved a substantial cost in terms of efficiency-seeking. By contrast, there are signs that Stewarts and Lloyds' caution on growth co-existed with stronger efforts at efficiency, suggesting a more equal balancing of the two pursuits. But the period of observation is too short and the evidence still too limited for any firm verification of the guiding hypotheses on managerial specialisation and long-term corporate phases outlined in Chapter 1. It is in the next chapter that some clearer confirmations on these begin to emerge.

### Notes

1 The following four paragraphs derive from D/DM, D/AGM and D/ACS; C. Wilson, 'Company Histories, I, Dorman Long' (c.1970); J. C. Carr and W. Taplin, *History of the British Steel Industry* (1962); C. A. Hempstead (ed.), *Cleveland Iron and Steel* (1979); and obituary notices on Arthur Dorman, *North-Eastern Daily Gazette*, 12.2.31, *The Times*, 13.2.31 and *Tees-side Chamber of Commerce Journal*, Feb. 1931.

2 D/DM generally. For Dorman's remuneration and share of profits see for example D/DM, 7.8.07. D/ACS, 6.12.1895, 11.12.1896, 4.12.1900, 11.9.01, 5.12.11, 10.12.12, 10.12.13.

3 D/ACS, 7.12.04, 7.12.09, and generally.

4 For Dorman's wartime remuneration and bonuses see D/DM, 13.2.17, 27.6.17, 12.11.18. For the Redcar project see D/DM, especially 11.1.16, 9.1.17, 13.2.17, 8.5.17, 13.7.17, 9.10.17, 12.3.18; D/AGM, 13.4.16 and 2.8.17. For the shipbuilding market euphoria during this period see also P. Payne, *Colvilles and the Scottish Steel Industry* (1979), pp. 146–50.

5 D/DM, 9.1.17, 17.4.17, 31.7.17, 13.11.17, 27.12.17, D/AGM, 20.12.17. D/CI (1066/5/12). Sir B. Samuelson and Co., Minute Book (1066/13/7).

6 D/DM, 6.1.20. D/AGM, 1.4.20. Carlton Iron Company, Minute Book (1003/19/5).

7 The Redcar project alone, with a planned capacity of 300,000–400,000 tons, represented an increase of between 46 per cent and 61 per cent on Dorman Long's total ingot capacity in 1913. During 1906–13 this had increased by only 55 per cent. For war profits see D/AGMs. For the new issues see D/DM, 13.11.17, 12.3.18, 10.12.18, 16.12.18, 11.3.19, 12.2.20, and D/AGM, 18.12.18, 1.4.20, 9.3.20.

8 D/DM, 8.7.19, 9.3.20, 7.9.20, 10.11.20, 14.12.20. During the four relatively good years 1911–14 dividends were distributed at an average rate of 7.5 per cent, during 1919–20 at 10 per cent. For the repayment assumption about the Redcar loan see D/DM, 12.3.18. For Sir Arthur's exuberance see D/DM, 1.4.20.

9 D/DM, 7.8.07, 9.3.15, 11.5.15, 30.9.15, 11.1.16, 11.5.16, 9.1.17, 17.4.17, 29.6.17, 13.2.18, 9.9.19, 6.1.20, 8.6.20. Papers on formation of Dormanstown Estate (1066/21/4). See also D/CI for examples of Sir Arthur's dominance of investment decisions.

10 D/DM, especially 11.1.16, 11.5.16, 9.1.17, 13.3.17, 17.5.17, 9.9.17, 13.11.17, 27.12.17, 13.2.18, 12.3.18.

11 D/ACS, 1892–1914. Author's interviews with retired Dorman Long officials. See also references in note (1) above.

12 P. W. S. Andrews and E. Brunner, *Capital Development in Steel: a Study of the United Steel Companies Ltd.* (1951). The author is indebted to Professor Elizabeth Brunner for access to data and interview notes used for this in

1950. Additional sources: G. R. Walshaw and C. A. J. Behrendt, The history of Appleby Frodingham (1950) (159/5/1); Steel, Peech and Tozer, General Minutes 1909–30; USC., miscellaneous historical papers, including predecessors' profits; Rother Vale Collieries, history from 1862 (159/5/2).

13   Andrews and Brunner, interview notes 1950. Obituaries on Harry Steel, *Sheffield Daily Telegraph*, 8.10.20, *West Cumberland News*, 9.10.20. Carr and Taplin, op. cit, p. 301. WISC, Final shareholders' meeting, 25.4.18. U/AGM, 28.1.19. Author's interviews with A. J. Peech and R. M. Peddie.

14   Sources for the next four paragraphs as in (12) above. For a criticism of the social efficiency of the mergers, arguing the locational disadvantages of Sheffield *vis-à-vis* cheap domestic iron ore supplies, see D. L. Burn, *The Economic History of Steelmaking 1867–1939* (1940), pp. 372–3. But see also Andrews and Brunner, op. cit, pp. 97–8.

15   J. Y. Lancaster and D. R. Wattleworth, *The Iron and Steel Industry of West Cumberland* (1977). Andrews and Brunner, op. cit. Andrews and Brunner, notes and interviews, 1950. Iron, Steel and Allied Trades Federation, statistical reports, 1915 and 1916. Workington Iron and Steel Company, Directors' Minutes and annual shareholders' meetings, 1908–18.

16   Sir J. Randles, letter to WISC shareholders, 21.3.18. Final meeting of WISC shareholders 25.4.18.

17   Andrews and Brunner, op. cit., p. 109.

18   Andrews and Brunner, op. cit., pp. 117, 122, 124–6. J. C. Carr and W. Taplin, op. cit., p. 326. U/DM, especially 27.1.19, 16.6.19, 18.8.19, 24.11.19, 23.2.30, 17.5.20, 12.7.20, 20.8.20, 4.10.20. For the scrap purchases see U/DM, 17.1.21.

19   U/DM, 19.4.18. Andrews and Brunner, op. cit., pp. 102, 105, 109, 116, 121. Andrews and Brunner, interview notes, 1950.

20   WISC Directors' Minutes, 27.9.17, 25.10.17, 10.11.17, 28.12.17, Sir J. Randles, letter to WISC shareholders, 21.3.18. Final meeting of WISC shareholders, 25.4.18. *Carlisle Journal*, 29.3.18. U/DM, 24.6.18.

21   Andrews and Brunner, op. cit., p. 121. WISC final shareholders' meeting, 25.4.18. U/DM, April 1918 to Oct. 1920. For Steel's industry-wide activities in 1918–20 see Chapter 9.

22   F. Scopes, *The Development of Corby Works* (1968) has useful pre-1914 material on Stewarts and Lloyds and its antecedent companies. See also Carr and Taplin, op. cit., pp. 261–2, 264, and Payne, op. cit., pp. 93–8. The following three paragraphs also draw on S and L records, chiefly S/CB (065/1/6), the British Tube Association's evidence in 1916 to the Departmental Committee, PRO/BT 55/39, and *Stewarts and Lloyds Ltd. 1903–53* (published by the company, 1953)

23   S/CB. S/BM. *Stewarts and Lloyds Ltd.*, op. cit. For boardroom reluctance in 1909 to take on new activities of which the directors were ignorant see Scopes, op. cit., p. 145.

24   Cited in H. W. Macrosty, *The Trust Movement in British Industry* (1907), pp. 46–7.

25   Scopes, op. cit., pp. 14, 16, 36, 138–47. S/BM. S/GPC, 1919–20.

26   Scopes, op. cit., pp. 16–18, 21–2, 81, 148–51. S/BM, 18.10.16. S/GPC, 13.6.19, 24.7.19, 29.8.19.

27   Scopes, op. cit., Chapters 1 and 2 for the background on LIC and Scottish steel supplies respectively. For the 1917 decisions involving NLIC see Scopes, op. cit., pp. 17–24 and S/BM, 18.1.16, 8.3.17, 4.7.17.

28   Scopes, op. cit., pp. 25–6, 29–30, 34–8, 152–5. S/BM, 29.1.20, 24.6.20, 25.8.20. S/GPC, 3.3.20, 30.3.20, 12.5.20, 15.9.20.

29   For acquisition projects, S/BM, 21.9.16, 21.11.18, 19.12.18, 29.1.20. For pricing policy, S/GPC, 13.6.19 and report to board, 23.7.19. For product diversification, S/GPC, 24.7.19, 24.11.20.

30  S/CB (particularly for valuable profit and other financial data); S/BM, 7.3.18, 3.12.10; S/AGM, 4.4.17. *The Times*, 5.4.18, 3.4.19, 6.8.20, 24.8.20.
31  S/CB. S/BM, 8.3.17, 7.3.18, 21.11.18, 13.3.19, 13.5.20 and 17.6.20. S/GPC, 12.3.19, 14.5.19, 15.9.20 and 24.11.20. S/OC, 23.7.19. Scopes, op. cit., p. 28.
32  WISC final shareholders' meeting, 25.4.18.
33  J. G. Stewart, 'The processes of manufacture of wrought iron and steel tubes', Institution of Engineers and Shipbuilders in Scotland, Glasgow, 1918, in S and L (1791/1/21).

# 4 Miseries and grandeurs of recession

This chapter broadly covers the period 1921–8, years dominated by economic recession and financial stress. A slump lasting from late 1920 until 1923 was succeeded by a brief semi-recovery in 1923–5 but then conditions grew worse. Competition from European countries revived, helped by postwar reconstruction, exchange rate conditions, government assistance, and other factors. In 1926 came the return to the Gold Standard and the General Strike. By the late 1920s many large companies were in a state of financial crisis. The government did little, providing neither effective tariff protection nor any active stimulus towards reorganisation. Not until 1928 did some faint signs of official or semi-official action emerge but by then, of course, it would take gargantuan measures from all sides to restore the industry to health, so far had the damage gone. Even then the slump had not reached its lowest point.[1]

Table 3.1 gave some idea of the depressing financial situation in our three companies. It must be cautiously interpreted because of the inter-firm accounting variations: for example, the USC profit figures are not fully comparable with the others. Several broad factors stand out. Stewarts and Lloyds' profits were both reasonable and relatively stable. Happily for this firm, too, profits continued to bear a reasonable relationship with the aggregate of issued share capital, debentures and long-term loans (only slightly over £4 million in 1921, about £5.5 million in 1928). By contrast, the USC's profits were low, on average, and varied the most. Also, they have to be set against a much heavier capital structure than Stewarts and Lloyds': on a broadly comparable basis, roughly £13 million in 1921, £15 million in 1928. Dorman Long's profits and returns on capital were also low. Moreover, this firm's position deteriorated the most steeply in so far as its declining profits have to be set against a large-scale increase during the period in its total of issued share capital, debentures and long-term loans: from about £9.6 million in 1921 to over £14 million by 1926. Both the USC and Dorman Long suffered from heavy fixed interest obligations in respect of heavy bank borrowing as well as debentures and long-term loans. Their depreciation provisions sank to abysmal levels. Dorman Long paid no dividends on ordin-

ary shares after 1921, the USC paid none on ordinary or preference shares after 1922.

Low profits are largely explained by economic forces and the
iron and steel industry's special vulnerability to recession (see
Chapter 2). The large investment projects of 1914–20, premised
on expanding markets and huge economies of scale, now became
liabilities in the main. The bitter fruits of the avoidable errors
committed during that period were now reaped. The conditions
for further capital investment were bleak, given cash shortages,
pessimism, high interest rates and capital market constraints. In
this situation it is tempting to regard the 1920s in British iron
and steel as a nadir of corporate power, a polar case of managerial
impotence. At this point, indeed, the lure of economic determinism could hardly be stronger. Yet the truth is that in such an
extremity management was still critically important, in some
senses more so than ever.

First, in terms of decision-making processes, the period provides a classic test of rationality in the face of turbulence. This
was partly a question of the attainability of high degrees of
clarity, consistency, logic and calculus under conditions of acute
discouragement and stress, partly a problem of whether managerial teamwork would strengthen or dissolve, of whether large
reserves of personal courage and resilience were available.
Second, the urgency of pursuing efficiency (of necessity, largely
outside production fields) called for a wide application of economic
concepts and management techniques. Just as we started to perceive the diversities of growthmanship in the last chapter, so in
this one we begin to compare managerial pursuits of efficiency in
some detail. The contrasts that emerge are no less striking. In
particular, Stewarts and Lloyds' marked efficiency edge over the
other two firms says much about implicit uses of management
techniques often questionably considered as 'modern'. Third, the
degree of *persistence* of basic managerial orientations towards
growth and/or efficiency becomes clearer. The survival of these
orientations through marked changes in the economic climate
was striking.

## Efficiency pursuits (Stewarts and Lloyds)

Of the three firms, Stewarts and Lloyds had the strongest board
during this period. Its older members included Henry Howard,
previously head of Lloyd and Lloyd, Brigadier General Hickman,
Robert M. Wilson, and the chairman, John G. Stewart, still in his
early sixties. Two other veterans, T. C. Stewart, the chairman's

brother, and James Menzies, had left the board in 1919, the latter after a major policy disagreement. Other dynastic figures and survivors of the 1903 merger included Albert Lloyd, J. H. (Jack) Lloyd and Joseph Howard, all relatively young and active. Representatives of a meritocratic tradition which, at least in the old Stewart company, dated back to the 1880s, included George Mitchell, a production man and longstanding manager of the Clydesdale Works, a director since 1907; Allan Macdiarmid, a young chartered accountant, secretary of the company, appointed to the board in 1918 at the age of 38; and Charles G. Atha, an iron and steel expert, appointed in 1920, who had a valuable background under Maximilian Mannaberg of the Frodingham Iron and Steel Company. Of the fourteen directors in 1921 only a minority were over 60; half were outside the original pre-1903 family circle of Stewarts, Lloyds and Howards; as many as five were recent recruits, and two of these, Macdiarmid and Atha, were extremely able.

The cause of efficiency was probably not spearheaded by John G. Stewart, who may indeed have been a conservative influence. But he must be given credit for bringing forward younger men and for at least tolerating substantial changes. Of the other older men neither Wilson nor Mitchell seems to have been a reactionary force. The principal younger star was Allan Macdiarmid who played a key role in the advance of financial controls and information systems and also, more importantly, in the evolution of thinking about general policy. His later eminence, though, must not be allowed to obscure the contributions of others. In particular, Albert Lloyd contributed influential ideas on raw material organisation and took a leading role in commercial policy and pricing decisions, as did Charles Atha on iron and steel issues. For example, Atha was mainly responsible for a 1922 report on ironworks co-ordination, which was a minor classic of its time on rationalisation, a prototype of many later ones both in Stewarts and Lloyds and elsewhere.[2]

In 1918 the board had set up a five-man General Purposes committee (GPC) 'to survey and focus the whole business', with particular emphasis on initial policy thinking, screening outside trends, planning 'future developments' and systematically formulating alternatives or recommendations to the board. This started a period of intense concern about the decision-making process. By 1920 the GPC was successfully pushing its views on the size and composition of the board itself. Efficiency, it thought, demanded that the board should go on being small. The idea of every sub-unit being directly represented tended to 'separatism and lack of co-ordination', endangering 'impartiality' and 'the full control over the various Departments that is so desirable'. The GPC's further proposals, for senior officials to receive higher status (in some cases through Local Directorships) and more say (through committee co-options), were implemented. Financial policy was specifically reserved for the GPC. Various committees covering overseas subsidiaries, office administration, allocation of orders and pricing were consolidated into a single Office Committee. A long standing division of the production side between a Works Committee (covering the tube works) and a Steel Works Committee was continued. But the last was to be 'a centralised co-ordinating authority' for 'all general questions connected with Coal, Ironstone, and Iron and Steel Works ... to make the best use of the present plant and to consider extensions as a whole in the interests of the Tube Works'.[3]

There can be little doubt that all this constituted a major drive towards a centralised, professional, functional organisation. In Chandlerian terms, this accorded well with Stewarts and Lloyds' specialisation and vertical character. The actual extent of centralisation, though, always hard to define and measure, was problematic. For one thing there was awareness of the penalties of 'a too rigid adherence to centralization'. In stores purchasing, for example, the board agreed it could lead to 'a lessening of the direct personal touch...and consequent decrease of interest and responsibility'. Also, there were the obstacles presented by traditional local and family interests. Thus, within the framework of a policy of moderate growth (see Chapter 3), it was relatively easy to pursue a Penrose-type absorption policy, pulling the new interests acquired since the war into the centralising current. It took only a few years to draw together the NLIC (acquired in 1917), Hickmans (1920) and the LIC (a Hickman subsidiary), and to subject them to the central organisation. But the 1903 merger heritage, which included some particularly resistant patches, was quite another matter. The long-established dual monarchy of the Scottish Stewarts, and the Midland Lloyds and Howards, presupposed continued elements of 'live-and-let-live' separatism and it co-existed with considerable degrees of local pride. Under the protective mantle of both traditions various anomalies and parochialisms persisted. For instance, not until 1919 and 1926 respectively were two long-established Clydeside outposts, the Vulcan and the Phoenix works, properly integrated within the group, the latter only because its manager died. Bigger bastions of each tradition were the Clydesdale iron and steel works in Scotland and the Coombs Wood tube works near Birmingham, both of them poorly-located, high-cost units, unamenable to central subjugation.[4]

Progress on the rationalisation of tube production was patchy. It was still dispersed among six works (two in England, four in Scotland), with total capacity divided about half-and-half between the two sides of the border. This structure could be reformed in piecemeal ways but not recast. Production was geared up for new market developments, for example for the fast-growing oil industry demand. A stream of technical improvements, often inspired by overseas visits, was introduced in one works or another: fuel economies; equipment for making weldless tubes; a continuous strip mill for butt-welding in one case, a large mill to replace five small, outdated ones in another. It was even possible to achieve some reorganisation within each national subgroup: in 1919–20 production facilities were redistributed between the two main Scottish works, Phoenix and Coatbridge, broadly concentrating ordinary tubes on the former, specialities on the latter. However, it was more difficult to shift the central allocation of orders from the existing rough-and-ready criteria towards a cost-minimising basis. Not until 1925, in the face of shrinking order books, was this attempted. As for a complete rationalisation, this would require large-scale finance, a linkage with iron and steel making and, almost certainly, a radical relocation move towards England, none of which was yet practicable.[5]

In other fields the pursuit of efficiency was more successful. Financial optimisation was vigorously attempted in a particularly sensitive area, that of internal raw material supply and transfer prices. This was a hornet's nest for the would-be rationaliser. On the one hand, there was a notion of corporate self-sufficiency, a concept of fully-employed iron and steel making facilities producing solely and adequately for the tube works' needs. On the other hand, there was a desire to minimise raw material costs or, more precisely, to maximise (long-run) group profits. These priorities could conflict for various reasons. Stewarts and Lloyds' iron and steel plant was not fully cost-efficient; particularly during a recession the tube works and iron and steel works might each sometimes do better for themselves by respectively buying and selling outside the group, to the possible detriment of the whole; and a local pursuit of self-sufficient intragroup transfers could sometimes conflict with optimisation when market transactions were, *per contra*, financially preferable. All at once these issues highlighted problems of economic logic, forecasting and human relations. Central optimisation policies were not only hard to formulate and keep up-to-date, they might also ride roughshod over sub-unit loyalties and morale.

To ensure priority for 'the interests of the Company as a whole' the board set guidelines for the works in 1923. Internal transfer prices of steel were not to exceed outside delivered prices 'plus perhaps a few shillings per ton to cover the advantages of continuity of supply'; the tube works should buy from the outside and the steelworks should sell externally only where this was clearly to the whole company's advantage; and to measure whether such exceptions were worthwhile the tube works should always procure minimal supplies from outside so as to keep 'an open door' and retain market price movements as a bench-mark. It was recognised that the policy might force a branch into loss-making 'in the best interests of the Company as a whole' – for example Hickmans were in that position in 1923. So the extent to which a works' financial results were distorted from 'those of normal trading' was to be estimated and reported.

This was a sophisticated policy and very much a financial optimiser's ideal. The guidelines cannot have been easy for the works to understand, let alone mutually agree upon, nor was it easy for the centre to pursue a pure quantified optimisation, even apart from the human relations problems. Difficult assumptions about future cost–output–market trends were unavoidable, so were frequent conflicts between short-run and long-run optimisation criteria. The board set up a committee of directors to act as a tribunal and appeal court for contentious cases, and the evidence suggests that this was active in implementing central *fiat*.[6]

Meantime there was marked progress in other directions. In 1917 a Stores Purchasing Committee was already doing useful work in 'regular consultation, combining requirements of various Works when purchasing, standardising where practicable the articles and qualities used'. Although by 1925 detailed enquiry revealed 'few important anomalies to be corrected', further improvements were made in group buying: the procuring of discounts on large orders, for example of oxygen, the consolidation and forward planning of coal purchases, and the testing of commonly purchased products. Costing systems were unified soon after the war. From 1919 onwards the control of stocks was strengthened, so was the control of working capital. Already by 1916 the works were being required to report to the centre on a wide range of matters and it is probable that their autonomy was considerably reduced over the whole period. The allocation of portfolio investment funds was an earnest affair. In 1922, for example, the GPC, not content with a handsome four-month capital gain of £80,000 on gilt-edged securities, demanded a careful analysis before making further investment decisions, 'looking to the probable course of the money and investment markets and the probable cash requirements of the Company' over a two-year period. On this occasion the General Manager of Lloyds Bank, Cooper Brothers, the accountants, and a firm of stockbrokers were all consulted. The keen eye for marginal gains was characteristic, so was the appetite for the best available internal and external information.[7]

In fact, it was in the development of information systems that Stewarts and Lloyds were most clearly in the vanguard. Here its methods reveal not only thorough fact-finding but also considerable conceptual clarity. The absence of explicit terms like value added, opportunity cost and time discounting did not exclude good approximations to the relevant economic concepts. Nor must the later semantics of management techniques obscure the fact that their basics could be substantially applied during this period by a relatively sophisticated management. Of course, the firm's system had faults. It reflected a strong financial emphasis and was probably less effective on the commercial side. In 1925, for example, we find the chairman complaining that overseas representatives' reports were lacking or else too narrow and statistical, whereas 'a free and full criticism of our trading methods, policy, quality of our goods and their suitability for the markets concerned...will be welcomed'. Inter-functional communications might also be weak, particularly between the production and sales staff. Even a minimal standardisation of financial information practices was patchy; easier, it seems, in Scotland than England. Capital project evaluations and profit estimates were still innocent of ranges of figures or explicit probability thinking. And there is no evidence that Stewarts and Lloyds' critical assumptions about economic prospects were any more refined than the general run, although they were generally more cautious and pessimistic, as before, which was, of course, no bad thing at this time.[8]

The following table lists a number of concepts and techniques, the date and occasion of their first recorded use and also, in some cases, how that use was described.

1   *Value added* (1913) Comparisons of purchases, production labour costs, overheads and finished goods sales values as between five Scottish works.
2   *Discounting* (1917) Estimated cost savings from two alternative acquisitions capitalised to 'present values', using alternative interest rates and time scales of capital redemption.
3   *Standard costing* (1919) Basing of uniform inter-works costing 'only on material, fuel, wages, maintenance and oncost, on the assumption that the Works are running full time.'
4   *Opportunity costs* (1919) Profit loss of nil action, *vis-à-vis*

continuance of an out-of-date works, 'estimated at about £70,000'. Exhaustive sets of alternatives, including selling and scrapping the works.

5 *Net cash flow forecasts* (1920) Dates of estimated returns, probable savings or increased efficiency, and expected payments on large capital outlay schemes, to be given in future.

6 *Incremental/marginal costs* (1922) Costing of new plant to concentrate on operating and variable charges, excluding head office charges. In 1925 this was refined also to exclude 'other works charges'.

7 *Cost–output functions* (1923) Statistics to be prepared 'shewing the effects of working at various stages of output varying from minimum to maximum' as a basis for deciding internal versus external raw material supplies.

8 *Long-term capital budgeting* (1923) Estimates of finance for schemes which might in the opinion of the various Managing Directors materialise during the next three years.

9 *Sub-unit shadow profits* (1923) Estimates to be attempted of hypothetical market-related works profits in addition to actual (guideline or headquarters-determined) ones, see above.

10 *Post-hoc evaluation of capital projects* (1924) Works to prepare detailed reports on a sample of past projects, 'showing whether or to what extent the results have justified the expenditure'.[9]

From limited production modernisation and rationalisation through central controls on buying, stocks, working capital, portfolio funds and transfer prices, across to greatly improved information systems, this was a formidable programme of reform. It made Stewarts and Lloyds a leader in the field of efficiency, although in some ways the USC's rationalisation and efficiency record in the 1930s was to be a greater achievement (see Chapter 7). It is an example of the generalisation that in many respects, though not all, recessions tend to favour efficiency pursuits. But it was, above all, a reflection of deep-seated managerial propensities. Efficiency was pursued by a managerial régime, well-established before the recession began, which had already shown signs of some bias in its favour (see Chapter 3).

It is worth adding that the Stewarts and Lloyds leadership also continued its keynote theme of caution, previously a bucking of the trend, now highly consonant with it. Major errors over the pace and financing of expansion were avoided. Despite anxieties over coal supplies there was no rush into colliery acquisitions. This minimised the financial–managerial quagmire such interests normally involved and probably served the firm well as coal prices slumped. The strategically important foothold in cheap Midland iron ores was quietly extended. So was the overseas position, with the development of tube-making subsidiaries in India and Australia between 1919 and 1921. Capital investment, mainly in tube-making plant, was quite heavy, averaging £312,000 in 1921–6 (compared with £107,000 during the war). Depreciation provisions were maintained at the firm's customarily high level, although accretions to reserves fell. Subscribed capital rose from £3.6 million in 1920 to £5.45 million in 1926. But, significantly, the extra finance took the form of deferred shares, not debentures whose 'dead weight' the company was still keen, and also able, to avoid.[10]

Whether the whole performance matched up to the astringent criteria of international competition or optimal economies is debatable. At least by 1925 the tougher-minded members of the Stewarts and Lloyds' board were convinced that it did not, and changes at the top soon meant that they would get their way (see Chapter 5). Meantime, however, the trend of Stewarts and Lloyds' development in the early 1920s should be clear. If the company was neither able nor willing to take bold forward steps, it at least avoided serious mistakes and kept its eye clearly on long-term growth strategy; and its overall record on efficiency was relatively good.

### Cliff-hanging (the USC)

The efficiency measures being taken in Stewarts and Lloyds were even more necessary and desirable for a group like the USC. A considerably larger size; a greater complexity in terms of producing units, products, markets and locations; more acute economic vulnerability because of a strong central focus on heavy iron and steel; the fact that the structure had been hastily and recently slung together: all these factors made it important for the USC to pursue internal reforms. Of course, the greater obstacles meant that measures comparable to Stewarts and Lloyds' would be more difficult to carry out but that would be no justification for neglecting them.

Instead, what occurred was a suspension of definite managerial phases during which neither growth nor efficiency nor social action nor any combination of them reigned supreme. It was not that growth, efficiency and social action propensities were totally lacking, but rather that none was strong enough to enforce his own, assuming he had them, and to some extent they deadlocked.

Nor was there a complete lack of managerial consensus, for this certainly existed in terms of the objective of sheer corporate survival. Nor did the growth–efficiency suspension necessarily exclude other types of managerial qualities, for example courage, resilience and ingenuity, in the cause of hanging on. A further point is perhaps more surprising. It is arguable that the nature of the USC's management between 1920 and 1928 was not necessarily the worst that could have happened for the firm. Although a determined pursuit of efficiency would clearly have been better, a continued passion for growthmanship during this period, on the lines of the last, would probably have been disastrous. That the USC *did* succeed in hanging on and avoiding certain extreme dangers therefore owed much, in a negative sense, to the suspension of strongly positive managerial pursuits.

Harry Steel, the combine's main initiator, lived just long enough to glimpse the lengthening shadows. The postwar boom was collapsing when he died prematurely in October 1920, the victim, it was said, of overwork. His successor as chairman, Albert Peech, also a son of one of the founders of SPT, had for many years been Steel's right-hand man. In all likelihood, the chairmanship devolved on him because of his seniority within the founding firm and his long closeness to Steel.

Albert Peech emerges as a thoroughly decent and assiduous chairman, not particularly bright, innovative or decisive.[11] It is hard not to feel sympathy for a man whose temperament, abilities and background scarcely equipped him for seven years' hard labour in a uniquely difficult job. There can be no doubt about Peech's allegiance to Harry Steel's concepts of expansion and vertical integration. Lacking an entrepreneurial flair for growthmanship, however, Peech's abilities fitted far better that part of the Steel legacy which conceived of the USC as a loose coalition of separate, although friendly interests. He showed no burning passion for efficiency and anyway lacked the astringency and techniques that would require.

What Peech did possess in abundant measure was doggedness. He soldiered on in the teeth of daunting odds. His concept of duty led to him shouldering an almost impossible combination of burdens: the continued managing directorship of SPT, the USC chairmanship, an active and assiduous role in the industry's national councils. On top of both operating and boardroom responsibilities within the main Sheffield–Scunthorpe–Workington axis of the USC, there were memberships of various committees of the national federation in London, the presidency of the federation in 1924–5, and frequent visits to Europe, for example twice in 1921, at least once in 1922, twice again in 1927. In summer 1925 Peech became seriously ill and had to have a long rest. Through the whole period there are signs of increasing fatigue and despondency (although, interestingly, Peech did not collapse into narrow attitudes on industrial organisation or labour matters: see Chapter 8). His private comments on the economic situation suggest frequent gloom, occasional exasperation: for example, 'One cannot help feeling that England must...come to the top again as a result of her having honestly paid her debts'; 'pessimism reigns supreme'; 'things are desperately bad in the whole kingdom'; 'I am absolutely at a loss to explain this state of affairs'. By 1927 Peech's characteristically frank comments to his immediate colleagues eloquently attest the weight of his cross: 'I am afraid you are having as miserable a time as I am', and, somewhat later, 'Well, I am not certain when it is all over I shall not be happier than I have been for a very long time'.[12]

How far was the USC board capable of making up for its chairman's limitations? It included several shadowy figures: J. E. Peech (Albert's brother), W. Tozer and Thomson Jowett, who were, like Peech, survivors of the old SPT regime, and F. E. Guedalla, a City man who gave some financial advice but, it seems, played no decisive role. Into a second category fell some personalities whose abilities, though strong in some ways, were probably narrow. Harry Steel's original partner in the mergers, Francis Scott Smith, continued as operating chief of Samuel Fox, whose interests he defended tenaciously. J. V. Ellis, the Workington managing director, was a conservative figure, inherited from the pre-merger régime, who represented his crisis-torn constituency with equal zeal. Neither of these men took a wide view of his role. That perspective was restricted to Sir Frederick Jones, Walter Benton Jones and Maximilian Mannaberg. As vice-chairman and as chairman of the finance committee, Sir Frederick Jones played a critical role. He was not simply an industrial statesman, an ex-chairman of the Mining Association who had represented the coal-owners nationally during important wartime negotiations with labour and received a baronetcy as a result. His long record as a successful operating head in the tough world of coal had also produced in him an overall corporate view and a sharp, if partially old-fashioned, attitude towards costs and profits. His background militated against growthmanship and adventurousness, in favour of ideas of financial caution and administrative control. Sir Frederick's son, Walter Benton Jones, was in charge of the USC's colliery operations and, although still relatively young, was both able and increasingly involved in strategic policy discussions.[13]

Yet only one man on the USC board possessed a commanding experience of the centralities of iron and steel, and that was Maximilian Mannaberg. Not that his great experience was able to be fully tapped. For although Mannaberg had retired from the Frodingham Iron and Steel Company in 1920, becoming a USC director in 1922, his energies were partly channelled into wider industry affairs. Probably more important still were conflicts of temperament and ideas. Unlike Sir Frederick Jones, Mannaberg seems to have been the sort of entrepreneurial chief executive who did not take kindly to corporate absorption. A dominating man with sweeping opinions and astringent ways of expressing them, Mannaberg cannot have been an easy colleague. He pressed strongly and consistently, if not always tactfully, for external caution and internal rationalisation.

The decision-making process that emerged from all this was consultative not only informally but also by intent. Albert Peech was neither keen to nor capable of exercising an individual dominance; the Joneses sought to complement rather than dislodge him; Mannaberg continued to be rather a 'loner', developing a strong entente with the Joneses but not with the chairman; and the quasi-feudal barons on the board, notably Scott Smith and Ellis, were powerful enough to exact important concessions to their sectional interests. An informal cabal of Peech and the two Joneses came to decide many strategy issues. Power was dispersed downwards through the hierarchy, to a large extent formally and deliberately. A finance committee under Sir Frederick Jones supervised cash movements and bank borrowing, and assessed capital investment proposals in *ad hoc* fashion; a small head office performed not much more than traditional accounting–legal functions; and a central committee of the board, largely comprising representatives of the sub-units, acted as a clearing-house, commentator and umpire between branches. The functional areas of sales, production, purchasing and research, possessed no clear representation on the board and little or no head office staff. Co-ordinated approaches there, if attempted at all, were left to a mixture of head office exhortation and (largely inter-branch) consultative committees. The relative autonomy of the sub-units was still further emphasised by the existence of works directing committees, effectively the old pre-merger company boards, vaguely mandated 'to direct the trade policy and maintain a general oversight of the activities of the subsidiary companies they are appointed to direct'. The creation of a largely consultative intermediate-layer works management committee in 1925, sandwiched between the central committee and the works directing committees, further complicated this already confused structure and probably achieved little in the way of co-ordination. It was all very confederal, rambling and diffuse, even, in a sense, democratic.[14]

During this period the foremost managerial activity was fire-fighting and crisis avoidance. The directors' minutes and, more important, the private papers reveal the many strains and stratagems involved: cuts in capital projects, stocks and current spending; anxious discussions about possible mergers and divestments (see below); efforts to avoid Excess Profits Duty liabilities; largely fruitless memoranda on such expedients as consolidating loan charges, raising second debentures or reshuffling subsidiaries for purposes of new share issues; resistance to auditors' suggestions that the assets should be substantially written down; dire warnings in late 1924 from the prestigious accountant, Sir Harry Peat; sensitive and occasionally tense relationships with the National Provincial Bank.[15]

However, from the viewpoint of our analysis, the important point is to consider how the USC's weak organisation combined with an underlying factor which, indeed, it partly reflected, namely a virtual absence of dominant constructive objectives. This void, in turn, was strikingly mixed in its results, on the one hand obstructing a determined pursuit of efficiency, on the other encouraging a useful caution which helped the combine to survive.

Of these tendencies the obstruction of efficiency is the more obvious. If Stewarts and Lloyds' production rationalisation during this period was patchy, the USC's hardly existed. A considerable number of small, dispersed and sub-efficient operating units was kept in being; the most blatant types of overlap continued, both in specialised lines like railway products and in bread-and-butter sectors like billets, with branches even sometimes directly competing in the same markets. Decisions on who was to fulfil which orders were a matter of horse-trading between branches at best, of catch-as-catch-can at worst. No determined drive was mounted on such matters as fuel efficiency. No long-term plan for production rationalisation against better times can be discerned. Meantime the pursuit of other forms of efficiency was also weak but with much less excuse. As with the production side, the full measure of the inadequacies in such areas as stock control, central purchasing, internal supplies, transfer prices, sales organisation and administrative overheads only emerged when rationalisation finally came after 1927 (see Chapter 7). Efforts there were half-hearted, fragmented and dispersed.[16]

The syndrome was reflected at the heart of the decision-making process: in the crudity and non-comparability of branch profit-and-loss data, the generally poor quality of written reports to the board, the long-drawn-out character of some deliberations, for example on a temporary stoppage of a works, the apparent inconclusiveness of some others, for instance on the division of orders between competing branches. With regard to the all-important deployment and renewal of managerial resources there is little evidence of top officials being retired or moved between branches or of outsiders being recruited either to operating units or to the headquarters.

Some sections, of course, showed a better effort than others. The management of the Frodingham iron and steel works and also of Rother Vale Collieries were probably relatively efficient. On the other hand, in the West Cumberland operations, whose financial results were, on average, the worst of any part of the USC, conservative attitudes dominated. The main indications were poor financial controls; 'rule of thumb methods' (Sir Frederick Jones's phrase); poorly argued investment proposals; approaches to stock control, cost efficiency and marketing which explicitly harked back to pre-1914 'normalcy'; not even modest efforts, it seems, to rationalise mining or production facilities stopping well short of (impracticable) major capital spending; and no steps to reform or renew the local management. True, the firm's maintenance of unprofitable activities in West Cumberland, as elsewhere, did not simply betoken inefficiency. Both economic expectations (hanging on till things got better) and social action (duties to dependent local communities) also played a part (see Chapter 8). True also, in a more overall sense, the inefficiencies could be worse elsewhere, say in a firm like Bolckow Vaughan. Nonetheless all these defects gravely weakened the USC at a critical time. There is no reason to doubt Mannaberg's contention, based on weighty experience and inside knowledge, that large sums could have been saved by better management (see Chapter 5).[17]

However, the suspension of positive managerial pursuits was not wholly bad for the USC. 'Organisational failure' stopped a long way short of disruption. Managerial vacillation was two-edged and even brought some benefits. It ruled out intemperate growth pursuits, inhibited bold ventures which could easily have come unstuck. If any powerful individual argued for risky moves, whether in the spirit of Harry Steel or in response to emergencies or through sheer naïveté, he could generally be over-ruled by others. Although the proponents of efficiency largely failed, at least they could often veto gratuitous threats to it. Reform might be vitiated by the overall stress on consensus and by conservatism in some quarters but, paradoxically, it was these same factors, strengthened by stoic personal qualities, which also helped to keep the whole ramshackle edifice intact. The USC might be stuck in a rut, even at risk of sliding downhill, but it was not to be allowed to disintegrate.

Sometimes a major error could occur, particularly if the matter appeared to be purely operational and within a branch's purview. Thus the temporary re-opening of the main Templeborough works in 1922, probably on the instance of Peech as SPT managing director, was largely responsible for the USC's serious losses in that year. But other mistakes, including some which could have been even more damaging, were avoided. Sir Frederick Jones, fretfully championing financial caution, kept an eagle eye on inter-branch cash movements and bank overdrafts, and was able to veto ill-thought-out proposals for new borrowing by Thomson Jowett in 1921, Maximilian Mannaberg in 1924 and 1925, and the company's accountants in 1923, on the grounds that these would further exacerbate the USC's already onerous debt burdens.[18] Successive discussions about takeovers or mergers illustrate even better the negative virtues of policy deadlock and shared control. It was the very pluralism of the régime which largely defeated the occasional tendency, tinged with desperation, to rush for panaceas in the shape of acquisitions, divestments or even absorption.

Thus when Guedalla and others advocated a takeover of the Steel Company of Scotland in 1921, largely in order to help the flagging Appleby plate-making project, this was knocked on the head by Albert Peech: the SCS had unwelcome problems and, anyway, 'I feel that we have sufficient on our hands at the present time'. A proposal to amalgamate Appleby with Richard Thomas's nearby Redbourne works in 1922 was turned down because of anxieties about both Redbourne's managerial defects and the risk of over-capitalisation of the resulting joint concern. Negotiations with the cash-laden South Durham Company, which sought to invest in or acquire both Appleby and Frodingham, were more difficult to stop because this time Albert Peech, by now 'obsessed with amalgamation', according to Walter Benton Jones, was in favour. The Joneses and Mannaberg strongly opposed this move, arguing the risks of over-capitalisation, loss of a still potentially lucrative asset, Frodingham, and a weakening of the USC in the event of an upturn in trade. By spring 1923 better economic prospects seemed to vindicate their argument as South Durham itself backed off the idea. Again, in early 1924 Peech was tempted by yet another merger notion, this time for a combination between Templeborough and other heavy steelworks in Sheffield. Walter Benton Jones headed him off with the view that the non-Sheffield parts of the USC would hardly benefit while more modern and therefore potentially more profitable plant than others in Sheffield would be lost: 'If we stick it out, we can swim while others sink.'[19]

## Victorian hangovers and fresh gambles (Dorman Long)

While managerial propensities worked well for Stewarts and Lloyds and their suspension brought mixed results to the USC, their operation during this period had, in the main, harmful implications for Dorman Long. Here previous strengths turned by and large into weaknesses and it was the very continuities of managerial behaviour which produced a severe dissonance with the economic environment and a heavy net accumulation of ills. The orientation towards rapid growth persisted long after its rationale had evaporated. A long-drawn-out and frequently painful retreat from the expansion syndrome threw off some bright sparks of endeavour, boldly and sometimes even successfully defying the surrounding gloom, but it also masked the realities of decline, postponed the necessary adjustments and threw up a further clutch of major errors which could arguably have been avoided.

Perhaps the most important single clue to the evolution of Dorman Long during this period, indeed up to 1931, is the fact that its now aged chairman, Sir Arthur Dorman, continued inexorably at the helm. The postwar expansion phase could have been an ideal moment for him to retire, still crowned with glory, but he did not do so, although by 1921 he was 74. His motives probably included a sense of duty, a lack of other interests, and a feeling of irreplaceability. But whatever the reasons, the implications seem reasonably clear. Sir Arthur showed few signs of relinquishing of power to others, exhibiting rather the well-known syndrome of a réfusal or inability to delegate. In a firm as large and ramified as Dorman Long this would have been a serious matter at the best of times. Now, following the major 1916–20 expansions and in the face of serious recession, major changes in both strategy and structure were all the more necessary. But it was unlikely that they would be introduced by a man so old and set in his ways. In fact, Sir Arthur showed many signs of clinging, naturally enough, to the patterns he had worked within for so long and, on the whole, so successfully: patterns of personal control and dynastic influence, patriarchal and imperial concepts of business, a short-cycle, upward-trend view of economic affairs moulded by his pre-1914 experience, and, not least, as we have seen, a pronounced growthmanship. This mixture of conservatism and expansionism was not without some benefits in the 1920s, just as Sir Arthur was still capable of remarkable achievement. But it was, on balance, inappropriate, just as he himself was now probably past his best.

The pattern was well reflected in Dorman Long's board. There Sir Arthur continued to be flanked by the still older figure of Sir Hugh Bell, a distinguished public man, inveterate speechmaker and indefatigable free-trader and Gladstonian Liberal. Although 77 by 1921 and a part-time director, Sir Hugh was still active. Non-executive directors were Francis Samuelson, ex-chairman of Samuelsons; Arthur Cooper and C. A. Head, also veteran survivors from previously independent firms now inside the group; Sir Edward Johnson-Ferguson, chairman of Bolckow Vaughan; and J. F. Mason, another senior figure from the political and business worlds. The marks of age and boardroom longevity, of previous acquisitions and political connections, were evident. Perhaps the most notable feature, however, was the inclusion of no less than eight members of the Dorman, Bell and Samuelson families (including Sir Arthur and Sir Hugh), a dynastic feat publicly eulogised by Sir Hugh in a characteristic invocation of the firm's 'family spirit' in 1923. The younger generation family element, by now predominantly middle-aged, was represented by four men on whom much hope appears to have rested, judging by their previous promotions to the board at young ages and their senior roles: Charles Dorman, Sir Arthur's elder son and heir apparent; Arthur Dorman, the younger and more assiduous son who was moving ahead of his elder brother by 1923; Maurice Bell, Sir Hugh's son and heir, who was involved in colliery matters; and Walter Johnson, a relation of the Bells whose father had also been a director of the firm.[20]

It is doubtful whether this inheritor element lent much strength to the board. To be fair, these men were probably constrained by a lack of outside industrial experience, the firm's rigid traditions and the dominating patriarchs, Sir Arthur and Sir Hugh, who tended to overshadow them. In the case of Charles Dorman and Maurice Bell the diversionary effect of outside (non-business) interests was, perhaps, unhelpful. There is no evidence of any capacity to champion radical changes or significantly to counteract Sir Arthur, let alone of the sort of abilities which would produce a good successor to him. Yet these men strongly outweighed the non-family executive members of the board. This, its smallest element, was represented by a retired soldier and general administrator, Colonel F. J. Byrne, head of the London office and a confidant of Sir Arthur whom, however, he was to predecease, and Laurence Ennis, a production man and internal promotee who joined the board in 1923 but remained outside its inner circle.[21]

Like the USC, Dorman Long was stuck with the legacy of the ambitious expansions of 1916–20. The largest of these, the Redcar project, which had already absorbed about £2.5 millions by the end of 1920, continued to take a lion's share of the much depleted investment funds available in 1921–2. Two other sizeable pre-1920 projects, improvements in the NESC plant and new steel sheet rolling mills, also had to be completed. This, combined with the miserable slump in profits from 1921 onwards (see Table 3.1) meant that large central stretches of Dorman Long's operations became virtually starved of funds. Much-needed investment in collieries diminished to a trickle. Coke ovens and blast furnaces, recognised as largely outdated, could not be replaced. One large and problematic steelworks, Clarence, could be neither scrapped nor properly improved, while another less problematic one, Britannia, could not be adequately modernised, so that by 1926 some of its plant was unsafe. During the years 1923–6 overall capital expenditure on new plant and equipment sank to a derisory annual average of £160,000. The backlog steadily increased so that by 1928, according to some (probably rather conservative) internal estimates, about £400,000 needed to be spent on the collieries and around £2 million on the ironworks, coke ovens and steelworks. The investment desert was not unique to Dorman Long, just as most of it lay outside management's control. On the other hand, management's responsibility for the situation was not negligible. This partly reflected some avoidable misjudgements in 1914–20 (see Chapter 3) but also a considerable amount of inefficiency in the 1920s without which the profits for investment would have been greater.[22]

The inefficiency issue needs to be interpreted in the context of the management situation at the top and Sir Arthur's long-established habits. In the early part of the period Sir Arthur seems ubiquitous as both chairman and managing director. Thus the company's chief chemist hurries to analyse 'a sample of stone taken by Sir Arthur at Kirby Knowle'; senior men anxiously consider how to get him to agree to a fifty-guinea fee for a mining survey in Morocco ('perhaps you may wish to speak to Sir Arthur about it when he is in better form'); and the directors hear him report verbally sometimes on this works, sometimes on that, sometimes on 'the operations of the various works', even sometimes on *all* of the works, mines and subsidiaries. In 1923 Sir Arthur handed over the managing directorship to his son, Arthur, perhaps partly reflecting the older man's commitments as president of the NFISM in 1923–4. Although Arthur Dorman took over operating responsibilities with regard to both production and trade representation, not only important policy issues but also current financial and other questions continued to be referred to the chairman right through the 1920s. He controlled negotiations for new capital issues, major investments and acquisitions, sometimes in conjunction with Sir Hugh. We also find him arranging the details of an overseas contract, allocating staff bonuses, selling investments, making overseas visits.[23]

Sir Arthur's age and personal rule, together with the board's lopsidedness, were probably bad for efficiency. For example, the absence of clearly defined responsibilities, articulated committees and clear reports to the board can hardly have encouraged proper control. While considerable *de facto* authority devolved on two particularly hard-working men, Colonel Byrne and the company secretary, T. D. H. Stubbs, others probably did too little. The question of who was in charge of finance remained particularly obscure. In 1923 the directors agreed to Sir Arthur's proposal for a finance committee, an idea which sank without trace. Charles Dorman had been made finance director in 1920 but some later evidence suggests that this arrangement was less than satisfactory (see Chapter 5). Various subsidiaries were not even formally integrated into the firm until 1923 and the 'Dormans versus Bells' tradition persisted. Overheads and administrative expenses were investigated sporadically with no clear result. A quasi-regal symbol of their embeddedness was the untouchability of Sir Arthur's own salary of £10,000 right through the decade. There are no signs of boardroom interest in rationalising purchasing, inter-works transfers, stocks, sales administration or clerical methods. And a hornet's nest of anomalies in financial information and control systems was later unearthed by the company's accountants, particularly from the late 1920s.[24]

It should be emphasised that Dorman Long was not necessarily falling behind in the field of technical ideas. For example, Arthur Dorman Jr showed a strong awareness of US and Continental developments in a wide-ranging survey of steel furnace size–efficiency relationships and of modern practices on open-hearth processes, mechanical handling, fuel efficiency and other matters, in 1925. An expert outsider's report on works practices in Cleveland in 1928 concluded that the firm was 'amenable to the consideration of ideas and contrasts noticeably with others in the district whose experience is more limited': a parochial triumph, perhaps, but not negligible.[25] This made the stagnation of investment all the more painful. But it also sets the organisational backwardness of the firm in even sharper relief.

If the overall cause of efficiency was ill-served, the banner of growth was still carried aloft. It is important to realise how Sir Arthur's perspective must have been largely shaped by his repeated experience of fighting his way through short-cycle economic vicissitudes since the 1880s. Some of these vicissitudes he had regarded as grisly at the time, notably those of the early 1890s. But at least six times recovery had come within two or three years at most, and the long-term trend for his firm had been upwards for more than four decades. So Sir Arthur was even less likely than younger men in the industry to anticipate over ten years of economic distress, all the more likely to remain fearless. As the 1920s wearily advanced, he continued to proclaim the old gospel of expansion, now, it is true, mingled with blood, sweat and tears, at the company's annual general meetings. Thus demand was picking up, production was to be maintained and further outlets sought (December 1922); the firm's tendency was 'to increase still further our output of raw materials and to extend the limits of our manufacture' (April 1923); difficult times lay ahead but 'it is to me as clear as daylight that [this] marks the end of Utopian programmes of shorter hours and restricted output' (December 1923); 'a more active trade must be built' mainly through increased labour effort and reduced prices (December 1924); 'I cannot help feeling that we are not far from the turning... Altogether, I cannot help feeling that we are nearing the end of the post-war depression' (December 1925); investment expenditures 'must come back to us at a profit' (December 1926); there had been 'a change for the better' and one could be optimistic about the prospects (December 1927). Not until 1927 did Sir Arthur proclaim cost-cutting efficiency measures as a major priority and even then there were side swipes at fashionable nostrums, including 'rationalization, whatever that may be'. It is doubtful whether any other steelmaster's growthmanship died so hard.[26]

Not that the expansion syndrome was totally without merits. True, an overseas venture could come badly unstuck, as did a minor one in Colombia. Further investment in the Kent coalfields in 1922 and a project for a steelworks there, involving the even more growth-minded Cowdray interests, turned badly sour. On the other hand, the company's stakes in South Africa and Australia were usefully developed, whilst new ones were added in the Argentine (1923) and India (1926–8), the former already proving profitable by 1926–7. After winning a celebrated £4½ million contract to build Sydney Harbour Bridge in early 1924, the firm set up a bridge-building department. Although the Sydney Bridge contract itself was destined to make a loss by 1931, the department as a whole proved a success. True, it diverted managerial resources from other priorities, further complicated the problems of controlling ramified operations, particularly overseas, and required still further supplies of strong nerves since the contracts normally took between two and four years to complete. But a stream of contracts both overseas and at home provided some desperately needed outlets for the firm's steel, profits built up to a respectable annual average of £40,000, and the counter-cyclical benefits were to become particularly evident by 1929–31. Arguably as important were certain less tangible benefits for Dorman Long. Contracts of over £½ million each to build bridges over the Tyne (1924) and the Thames at Lambeth (1928) made its operations strikingly visible to the ordinary public. An increasing roll-call of overseas contracts, for example in connection with the Nile Bridge (1925) and the Beira Railway Company (1927), provided an impeccable patriotic underpinning and wrapped the firm still more in the British flag. During a bleak period there were probably considerable benefits from all this in terms of bank credit, capital market sentiment, government and public goodwill, and employee morale.[27]

But the growth syndrome's darker side lay particularly in its encouragement of financial errors. It was in the field of financial policy that the firm's (and more especially Sir Arthur's) biases exacted perhaps their heaviest toll.

Dorman Long had resorted extensively to debenture borrowing before the war. Its capital structure, like the USC's, became heavily inflated as a result of massive investments and acquisitions in 1916–20, leading particularly to further long-term borrowing. Again like the USC, Dorman Long experienced this as a heavy burden as the recession worsened. However, *unlike* the USC, Dorman Long proceeded to exacerbate the situation. Seeing the mild recovery of early 1923 as a critical turning-point, the firm made further issues of preference shares and, more significantly, mounted a £3½ million 5½ per cent mortgage debenture issue in April that year. This was intended not merely to pay off a government loan inherited from the war and to reduce bank borrowing but also to finance large-scale expansion. The issue was exceptionally large relative to others recently in the industry: absolutely, as a proportion of paid-up capital or capital employed, and in terms of the resulting gearing ratio. As a result Dorman Long's capitalisation was increased to about £12 million, soon to be pushed up still further by another debenture issue of £½ million in late 1926. By 1927 the firm's total indebtedness was over £7 million (£5 million of debentures, over £2 millions of bank loans).[28]

There can be little doubt that the central decision in this process, the 1923 £3½ million debenture issue, reflected a belief that the 1923 upturn heralded a genuine boom, a belief coloured by decades of pre-1914 cyclical experiences and fortified by Sir Arthur's expansive impulses. If he was not alone in making that assumption, his long-established growthmanship meant that he was prepared to place much higher bets on it than others were. In a way it was a final grand eruption of Victorian exuberance, a last gambler's throw. The result was a more rapid slide towards financial weakness.

## Conclusions

All three firms began this period with a major backlog of problems related to efficiency. The large-scale mergers and acquisitions inherited from 1914–20, even from the turn of the century, had still by 1920 not been efficiently absorbed in the main, and the organisational efficiency problems of both size and vertical complexity had remained largely unaddressed. These lags of efficiency behind growth were already serious even before the onset of an economic recession which soon perverted the bold scale-economy investment projects of 1914–20 into millstones, exacerbators rather than improvers of unit costs. Thus three types of efficiency problem were superimposed, those relating to merger-led growth, size and verticalisation, and unpredicted economic crisis. Although their severity differed as between the firms, the logic of the situation was broadly the same.

The resultant challenge to managerial efficiency-seeking was enormous (although, it should be added, not necessarily more complex than the sensitive issues of centralisation, decentralisation and efficiency maintenance during prosperity which emerged later: see Chapter 7). Since large-scale capital investment was financially impracticable, the challenge necessarily arose largely outside the field of production rationalisation. The responses of the three firms varied strikingly, as we have seen. Stewarts and Lloyds set the pace particularly in terms of the shape of its top decision-making unit, the breadth, timeliness and quality of its internal and external information flows, the economic concepts which governed both the form of these flows and their interpretation, and the parallel pursuit of a centralising rationalisation mainly in the non-production aspects of the organisation. It is clear that both the other firms lagged seriously behind in all these ways and that, largely as a result, they were all the more destined for trouble.

From these contrasts, superimposed on those of the last chapter, a clearer picture of the operative managerial–corporate forces begins to emerge. Much of the typology of corporate long-run policy phases suggested at the beginning of this book gathers sustenance as we observe striking changes in conditions and boardroom responses to them. The suggested characterisation of these policy phases in terms of a growth concentration, an efficiency concentration, a growth–efficiency balancing act or indeterminacy is seen as a fair approximation, a reasonable depiction of central trends (the question of social action will be considered later). Thus in the case of Stewarts and Lloyds a balanced combination of efficiency and growth-seeking, glimpsed in the last chapter, emerges more clearly. In the USC the growth concentration characteristic of the Harry Steel era is succeeded by a phase of suspension of positive pursuits. In Dorman Long the long phase of growthmanship primarily associated with Sir Arthur Dorman is shown to have strikingly persisted.

Of course, these are simplifications and we have certainly seen something of the rich diversity of the relevant managerial factors: for example, the power sharings in both Stewarts and Lloyds and the USC as compared with Dorman Long's autocracy, the inherited values of sober judiciousness in Stewarts and Lloyds, of libertarian confederalism in the USC, of Victorian exuberance in Dorman Long. We have even glimpsed some of the individual psychological subtleties, for instance Albert Peech's mixture of bewilderment, vacillation and doggedness, Sir Arthur Dorman's obstinate courage and reliving of past battles. But we have also seen that these diversities tended broadly to cohere around a few central trends. The heritages of previous phases, the power distributions at the top, and the values and abilities of the leading individuals arguably contributed, for all their variations, to clear policy constellations: a careful balancing act, a survival-clinging hiatus, a growth bias. It was these policy orientations, superimposed on economic conditions, which respectively helped Stewarts and Lloyds to fortify itself, the USC to stay confused but still together, and Dorman Long to lurch towards severe internal disorders.

Not only did the policy concentrations take their colour from managerial–corporate factors rather than changing economic conditions. They also showed a marked resilience throughout these changes. Considerable tenacity and courage were evident. None of the managerial régimes cracked under the strain. None responded ultra-sensitively to the recession by revolutionising itself from within or having a change of top personnel externally imposed. Even where a corporate emphasis married reasonably happily with the new economic circumstances, as in Stewarts and

Lloyds, that emphasis was an inheritance from previous years. Where a change of régime occurred, the USC case in late 1920, this owed nothing to the recession and indeed was substantially ill-adjusted to it. As for Dorman Long's continued growth-manship, this represented an outright defiance of the new context in many significant ways. Thus exogenous pressures modified the operation of the fundamental managerial biases and specialisations but did not alter them.

## Notes

1  For postwar and 1920s economic conditions in the industry see particularly J. C. Carr and W. Taplin, *History of the British Steel Industry* (1962), pp. 346–80, and D. L. Burn, *The Economic History of Steelmaking 1867–1939* (1940), pp. 395–428.

2  S/CB. Miscellaneous material on committees (1791/1/14). S/BM. S/GPC. P. Payne, *Colvilles and the Scottish Steel Industry* (1979), p. 92. Author's interviews with ex S and L officials. F. Scopes, *The Development of Corby Works* (1969), pp. 30, 38–9, 156–61.

3  S/BM, 21.11.18. S/GPC, 14.5.19, 28.1.20, 3.3.20, 3.11.20.

4  S/BM, 3.5.17. S/OC, 10.3.21. Scopes, op. cit., p. 30. Author's interviews with ex S and L officials. S/GPC, 13.6.19, 24.7.19, 22.10.19, 26.10.21, A. C. Macdiarmid, 15.12.19 in S/EC.

5  Oil tubing and technical changes generally are outlined in Notes on recent developments for staff, 1924 (1791/1/23). Fuel economies were mooted in 1919: S/GPC, 13.6.19 and S/WC, 3.12.19. For the limited production rationalisations see S/GPC, 14.5.19, 13.6.19, and Scopes, op. cit., p. 159. For order allocation problems see S/GPC, 28.1.20, 4.6.25.

6  S/WC, 14.5.19, 12.6.19, 8.8.19, 27.1.21. S/BM, 6.9.23, 18.10.23. Materials Price Regulating Committee, 16.11.23 and 11.12.23 in S/EC. Secretary's Department to R. Smith, 31.7.25, S/EC.

7  S/BM, 3.5.17. Stores Buying Committee, minutes, 1925 and 1926. S/WC, 24.10.18, 29.1.20; Secretary's Department to R. Smith, 14.3.22 and further correspondence on costing system unification, July 1924, in S/EC. S/WC, 22.11.16, 31.5.17; A. C. Macdiarmid to Works Managers, 17.2.20 in S/EC; S/OC; 23.7.19. S/GPC, 13.6.19, 12.5.20, 2.5.22, 21.6.22.

8  S/GPC, 7.5.25. S/OC, 17.7.25. In 1932 an Economy Commission found that 'fairly radical changes' were needed in costing systems in the English works but not in Scotland (1791/1/21). On single figure estimates, 'an impossibly accurate forecast of the rate of profit' was abjured in S/GPC, 24.11.20.

9  (1) Letter to A. C. Macdiarmid, 28.5.13 in file on tube prices (1791/1/8). (2) Statement by G. A. Mitchell on relative values to S and L of NLIC and LIC, in Scopes, op. cit., p. 148–51. (3) S/GPC, 13.6.19 (4) S/GPC, 24.7.19. (5) Net cash flow forecasts 1920 origin. (6) Secretary's Department to R. Smith, 13.3.22 in S/EC, and S/GPC, 4.6.25. (7) Materials Price Regulating Committee, 6.3.23, in S/EC. (8) S/BM, 14.6.23. (9) Materials Price Regulating Committee, 16.11.23, in S/EC. (10) S/GPC, 15.24.

10  S/BM 13.3.19, 17.6.20, 21.1.21, 10.3.21, 8.9.21, 10.3.22, 30.3.22, 3.5.22, 30.11.22, 12.1.23, 9.3.23, 28.3.23, 27.3.23. S/GPC, 5.2.19, 24.11.20. S/CB. S/AGM, 20.8.24. Scopes, op cit., pp. 47, 49–50. The board was prepared to pay £750,000 for one colliery in a deal that fell through in early 1923, but purchased another for £157,500 later that year.

11   The following judgements on Peech mainly rely on the archival material, to a
     lesser extent on interviews with retired USC officials.
12   U/DM, especially 9.11.20, 17.1.21, 15.2.21, 16.8.21, 24.1.22, 8.7.23, 5.8.24,
     9.9.24, 4.11.24, 2.12.24, 3.2.25, 10.3.25, 12.5.25, 9.6.25, 7.7.25, 8.9.25, 6.10.25,
     10.11.25, 8.12.25, 9.3.26, 8.2.27, 10.5.29. U/CC, 4.6.25. A. O. Peech to H.
     Guedalla, 8.7.21, A. O. Peech to Sir F. Jones, 4.5.27, in Amalgamation files
     (006/10/3). W. Benton Jones to Sir F. Jones, 23.12.27 in U/RA (006/24/37).
13   U/DM, U/CC, U/FJ, U/WWD, Amalgamation files (006/10/3) and Miscel-
     laneous historical papers.
14   U/DM, especially 9.11.20, 15.12.20, 13.9.21. U/FC. U/CC, especially 12.9.22,
     10.10.22, 3.3.25. U/RA, Andrews and Brunner, interviews, 1950.
15   U/DM, especially 17.1.21. 15.2.21, 15.3.21, 23.5.21, 15.11.21, 16.12.21,
     24.1.22, 5.8.24, 7.9.24, 2.12.24, 12.1.26, 13.7.26. U/FC, 1.12.20, 14.12.20,
     11.2.21, W. D. Scrimgeour to Sir F. Jones, 25.1.24. U/FC/C. General Matter
     file (006/10/3). U/FJ. Amalgamation files (006/10/3). Steel, Peech and Tozer,
     General Minutes, 1923–5.
16   Tyres and axles were manufactured in both Workington and Sheffield, see
     U/CC, 4.9.28. Rails were made at three separate works, see U/RA. For
     inter-works competition on billets see U/RA, 31.10.27. For order allocation
     problems see particularly U/CC, 7.3.22, 27.3.22, 12.8.22. For the slow start of
     fuel efficiency measures in 1923–4 see U/CC, 20.6.23, 28.8.23, 1.7.24. On
     inter-works supplies, a comment by Peech was characteristic: these 'should
     receive sympathetic consideration from all concerned', U/DM, 13.9.21.
17   U/FJ. U/WWD, 28.2.21, 31.3.21, 11.5.21, 7.6.21, 1.11.21, 6.12.21, 12.12.22,
     6.2.23, 13.3.23, 27.9.23, 21.1.24, 29.1.24, 2.2.26, 28.9.26, 30.11.26, 4.1.27,
     1.3.27, 5.4.27, 31.5.27, 5.8.27, 29.11.27. U/FC, 11.3.21, 12.4.21, 26.10.21,
     3.11.21, 24.11.21, 2.2.22, Sir F. Jones to M. Mannaberg, 12.8.26. U/DM,
     11.10.21, 16.12.21. U/CC. 6.12.21.
18   U/DM, 16.5.22. U/AGM, 10.10.22. U/FC, 11.2.21; Sir F. Jones memorandum,
     10.12.23; M. Mannaberg to W. Benton Jones, 6.12.24 and 4.1.25.
19   Amalgamation with various concerns 1921–3 (006/10/3). U/FC/C. U/FC,
     14.12.20. U/DM, 27.6.22, 25.7.22, 29.8.22. U/FJ.
20   D/DM. 'Who was Who'. Samuelson, born in 1861, an inheritor within his own
     firm and a Dorman Long director since 1911, often missed board meetings on
     account of ill health. Mason, born in 1861, chairman or director of several
     other companies and an important NES Co shareholder, had been a director
     since 1903. Criticised for his poor attendance at the 1905 AGM, he seems to
     have been very inactive but continued on the board through the 1920s.
     Johnson-Ferguson incurred much criticism as the man who presided over
     Bolckow Vaughan's rapid decline, which climaxed in 1924 (see Chapter 5).
     For Sir Hugh's eulogy see D/AGM, 17.4.23. Charles Dorman (1875–1929)
     became a director in 1902 and acted as deputy chairman in the 1920s;
     Maurice Bell (1871–1944) also joined the board in 1902; Arthur Dorman
     (1881–1957) did so in 1908. The two Dorman sons were joint managing
     directors in the 1920s. In early 1920 their commissions on profits were fixed
     at 1½ per cent each as compared with Sir Arthur's 3 per cent, see D/DM,
     12.2.20.
21   These judgements depend mainly on the documentary evidence, partly on
     interviews with retired officials. Charles Dorman had been a Middlesbrough
     town councillor, Mayor in 1903 and Conservative Parliamentary Candidate
     in 1910. He was a JP, President of the Middlesbrough Conservative Associa-
     tion until 1928, and Joint Master of the Cleveland Hunt. See obituary,
     *Northern Echo*, 1.3.29. Maurice Bell had been a professional soldier in the
     Boer War and the First World War, was High Sheriff of Durham in 1921, and
     was also very interested in sporting activities.

22 D/DM, D/CC, D/FF. In correspondence with Whitehall in 1930–1 Dorman Long's coke ovens were described as 'largely obsolete' by one of its directors. But H. A. Brassert told Sir Horace Wilson that, even with new coke ovens, the blast furnaces were insufficiently up-to-date to take advantage of an improved gas supply. See PRO/BT 56/37, CIA 1093/9, 25.11.30, 3.7.31.

23 Chief Chemist's letter books (203/1/77), 30.4.20. D/Cl (1066/5/4), F. J. Byrne to T. D. H. Stubbs, 16.12.21, 23.13.31. D/DM. Sir Arthur was the key figure in the 1923 debenture issue, the 1927–8 merger negotiations with Bolckow Vaughan, and overseas operating decisions, for example about Australia in 1927–8. He visited Germany in 1923, the Argentine and Australia in 1924, South Africa in 1928. For his close involvement in detailed financial affairs in 1927–8, see Chapter 5.

24 Confused responsibilities for finance are manifest in the differing personnel reporting on it to the board even after Charles Dorman's appointment as finance director in July 1920: sometimes the latter, sometimes Sir Arthur, sometimes the Secretary, occasionally an unattributed report. For overheads see D/DM, 22.7.26, 14.9.26. A 1928 memorandum by the Chief Accountant referred to 'some very extraordinary figures' for one works, including goodwill and suspense items dated 1899, see D/FF. Interview with Mr J. Jack. For the major rationalisations still needed by the mid-1930s see Chapter 7.

25 Arthur Dorman, presidential address to the Cleveland Institution of Engineers, Proceedings 1925–6, 17.10.25. P. H. Andrews, Report on Teesside Blast Furnaces 1928 (unpublished). I am grateful to Dr J. K. Almond for lending me the latter.

26 D/ACS, particularly 5.12.1894. D/AGMs, 15.12.22, 18.5.23, 24.12.23, 11.12.24, 17.12.25, 16.12.26, 8.12.27.

27 D/Cl (1066/5/3, 4, 12 and 14). D/DM, especially 12.5.21, 12.9.23, 9.10.23, 13.11.23, 12.5.24, 12.5.25, 12.5.28, 16.2.26, 13.5.26, 17.1.28, 20.3.28, 15.5.28, 19.6.28. D/FF. Sydney Harbour Bridge Contract (1066/8/3) Bridge Department (1066/8/4). A 1933 survey showed that, on 28 bridge contracts since 1925, the aggregate profit had been 7.5 per cent, the upper quartile profit 11.9 per cent, the lower quartile profit 1.0 per cent. Seven contracts produced losses. The 14 overseas contracts were more variable in profitability and less profitable overall than the 14 UK ones (1066/8/4). A further survey in 1937 referred to the advantages to the steel mills and the constructional department during the depression, also to 'various unseen benefits' (00313).

28 D/DM, 27.3.23. D/FF (1066/9/1, 2 and 3). *The Times*, 10.12.14, 18.4.23, 11.5.23, 13.5.23. The following table compares Dorman Long's 1923 debenture issue with others by iron and steel companies between 1918 and 1923. Source: *Times Books of New Issues*.

Table 4.1. *Debenture issues by iron and steel companies*

| Company | Year | Amount (£ 000s) | Net asset cover, times | Past periods, average profit cover, times* |
|---|---|---|---|---|
| Ebbw Vale | 1918 | 600 | nearly 5 | not given |
| Barrow Hematite | 1919 | 600 | n/a | not given |
| Millom and Askam | 1919 | 1000 | n/a | not given |
| Beardmore | 1920 | 1000 | over 5 | not given |
| Ebbw Vale | 1920 | 1500 | over 4 | not given |
| Hadfields | 1920 | 1000 | nearly 2½ | nearly 5 |
| Baldwins | 1921 | 2250 | over 2½ | over 3½ |
| Pearson and Knowles | 1921 | 1000 | over 3 | 4½ |

Table 4.1.   *continued*

| Company | Year | Amount (£ 000s) | Net asset cover, times | Past periods, average profit cover, times* |
|---|---|---|---|---|
| Richard Thomas | 1922 | 1000 | 2⅓ | 6½ |
| Consett | 1922 | 1500 | 3½ | over 5 |
| Bolckow Vaughan | 1923 | 1000 | nearly 2½ | over 3½ |
| Dorman Long | 1923 | 3500 | 2½ | over 3 |

* Related to average profits for previous periods of between 4 and 13 years

It will be seen that Dorman Long's issue was exceptionally large in absolute terms, put the firm among those with the highest ratios of debentures to net assets, and, judging by average recent profits, appeared to have the lowest prospective earnings cover for interest and repayment. In addition, Dorman Long's bank borrowing may have been above-average.

# 5   The managerial watershed

This chapter considers major changes which occurred in the top management personnel of the three companies: in Stewarts and Lloyds in 1925–6, in the USC in 1927–8, in Dorman Long in 1931. As the three firms moved towards these boardroom transformations they all exhibited, sometimes more vividly than before, the syndromes of the early 1920s; Stewarts and Lloyds' efficiency-plus-growth orientation, the USC's managerial hiatus and Dorman Long's growthmanship hangover. In all three firms it is arguable that the changes themselves critically depended on managerial–corporate forces. External economic influences were important but far from decisive, let alone supreme. However, perhaps the most important feature of this chapter is that it is possible to examine the values and abilities of the men who took over more fully than was the case with their predecessors, including, in most cases, their stances on social action as well as on growth and efficiency.

## A natural succession

Stewarts and Lloyds' managerial changes occurred first, involved the most continuity, and are the easiest to interpret. Neither their timeliness nor their smoothness should come as a surprise. They largely reflected the firm's efficiency pursuits, its relative ease in riding out the recession and its reserves of managerial talent (see Chapter 4).

In March 1925, after a period of failing health, the elderly chairman, James G. Stewart, died. In retrospect it is clear that his most obvious successor was Allan Macdiarmid. The reasons why the board failed to appoint Macdiarmid straightaway were probably mainly dynastic and conservative. The tradition of a chairman from one of the founding families probably died hard, and some leading figures in the firm may also have had apprehensions about Macdiarmid's sharpness, ambition and zeal for rationalisation. At any rate the decision was to appoint Robert M. Wilson, an elder statesman, probably on an explicit caretaker basis, and to make Macdiarmid joint deputy chairman.[1]

The Wilson–Macdiarmid régime took over at a time of corporate stress. Pre-depreciation profits fell from £579,000 in 1924 to £366,000 in 1925 and reached their lowest interwar point in 1926 at £262,000. But although serious, this was far from calamitous. Ample reserves, low levels of bank borrowing (still only £½ million in summer 1926), a virtual absence of debentures and a continued ability to pay dividends, albeit much reduced, shielded the firm from the extreme humiliations of some others. This is not to deny that lower profits and drastic reductions in capital investment (cut from £538,000 in 1924 to £266,000 in 1925 and £72,000 in 1926) caused much heartburn. Indeed, these stresses helped to justify the new régime in carrying out some major changes in 1925 and 1926. First came the incorporation of a liquidated Alfred Hickman into Stewarts and Lloyds, which led to opposition and resignations, followed by a searching examination into Hickmans' future as a high-cost operation, leading to its eventual reprieve. Second, various economy measures were taken. Third, a consultant's report led to far-reaching changes in the management of the still trouble-ridden NLIC. Then, in November 1925, Robert Wilson asked the aged joint deputy chairman, Henry Howard, to retire (he had been in Lloyd and Lloyd since 1871). The board underlined this by emphasising the old man's departure also 'from all participation in the affairs of the company'. These decisive but far from comprehensive acts of rationalisation probably helped to clear the way for Macdiarmid's elevation to the chairmanship which finally took place in June 1926.[2]

We must now consider in some detail the personality of Allan Macdiarmid, chairman of Stewarts and Lloyds from 1926 to 1945. In some ways his characteristics have been obscured by a species of hero worship, implicitly in the case of an 'inside' history by his protegé, Sir Frederick Scopes, explicitly on the part of the economic historian, Duncan Burn.[3] An attempt to get at the truth about Macdiarmid must start out from the facts of his early life.[4] He was born in 1880, the son of a Glasgow merchant, into a family with Highland forebears but long settled in Glasgow. Baptist influences were strong. Macdiarmid remained interested in religious issues but later cut loose from denominational ties and brought up his children as free-thinkers. After attending Kelvinside Academy in Glasgow, he proceeded to a minor public school, Uppingham. Although athletic and a good games player, he already showed a marked propensity for the arts: literature, drawing, painting and more particularly music. He is said to have contemplated a musical career at one point. Later, as a successful businessman, he was to intersperse takeover negotiations with piano playing and to obtain much private refreshment from playing Beethoven.

However, a stark division between the private man and the public achiever was already evident in Macdiarmid's choice of career. He trained as a chartered accountant in the Glasgow firm of M'Clelland, Ker. In 1909 he entered Stewarts and Lloyds as company secretary and at about this time, too, he married. His wife, a Glaswegian of similar background, was apparently a woman of strong character and rigorous views. By 1913 Macdiarmid was receiving confidential notes addressed to the chairman of Stewarts and Lloyds; by 1918 he was a leading member of the firm's inner cabinet; by 1924 he was flanking the chairman in outside negotiations and was entrusted, for example, with the 'supremely important task' of 'strengthening the Selling Department'. This rapid ascent, a tribute to Macdiarmid's industry, intellect and determination, was also a signal achievement for an accountant and an outsider, and in wider terms, too, a triumph of managerial professionalism.

Most of the evidence on Macdiarmid's business ideas is contained in some surviving correspondence, numerous papers to the board, his distinctive acts as chairman and, less importantly, his public speeches. To begin with, there was Macdiarmid's great debt to the corporate traditions of the firm. This, rather than the reverse typology favoured by the 'great man' approach, is an essential clue to his achievement. Hardly a radical innovator, he was steeped in the long-established style and ethos of Stewarts and Lloyds, its reasoned argumentation about policy, its stress on prudent financial conservation, its consistent long-term goal of supremacy in the steel tube trade, its strategy of backwards vertical integration. He must have learnt much from his apprenticeship under older men like James Graham Stewart. Even the Corby project in the 1930s, closely identified with Macdiarmid personally, owed much to earlier corporate interests both in Northamptonshire and in the ideal of a truly cost-efficient iron and steel operation in the service of tube production.

What Macdiarmid brought to these largely established pursuits was impressive intellectual and practical ability and exceptional flair. We can only speculate as to the deeper sources of his determination to succeed. Certainly, desire for money, titles or social status, even for the normal public appurtenances of power, hardly featured. More evident were an intellectual fascination with business, an overarching search for order, a quiet zest for the power and competitive game, an urge to do technical jobs well. Macdiarmid's ambition was so fused with the firm's that personal and corporate interests were inseparable to a particularly high degree. With this dedication to the advancement of Stewarts and Lloyds neither sectional loyalties nor civic links nor national

affiliations were to be allowed to interfere. Macdiarmid's strength
inside the firm, both before and after 1926, partly reflected his
detachment from its internal cross-currents, his ability to arbi-
trate and look ahead in the overall corporate interest. In this
sense he benefited from non-membership of the founding dynas-
ties and also from a refusal to let his own Scottish and Glaswegian roots stand in the way.

Macdiarmid's papers to the board display exceptional intellectual ability and lucidity of expression. Characteristically, he
spent much time preparing them, getting every detail right, even
though his colleagues, deferential to his authority, would probably have accepted much less. Here and in his speeches the
language is logical, rational and spare, devoid of emotion or
moral overtones. Only the occasional military metaphor, the
strategic analogies of 'offence' and 'defence', is used. Even the
set-piece annual chairman's speeches are devoid of padding and
rhetoric, with the rare colourful word harnessed to some functional pursuit: for example, the 'sacrosanctity' of the home market, a
'constant dread' of obsolescence. The reticence and economy were
evident in other ways. Macdiarmid's correspondence with other
businessmen was a model of elusive carefulness; in industry-wide
deliberations he engaged in formidable silences; in discussions
with major figures like Montagu Norman he was adept in conveying a tone of non-committal co-operativeness; in important
negotiations he was accompanied by an amanuensis. Equally if
not more important were persistence and an ability to play things
long, a *penchant* for provoking argument among his immediate
colleagues in the interests of informed decision-making, an ability to attract their loyalty, an incisive power to command.

Of course, Macdiarmid's brilliance had limits. To a large extent
these reflected the marked split dividing his business ideas and
activities from his personal life. Whatever the cause of that division, it had several clear effects. If his personal life included
elements of warmth and humour, Celtic romanticism, literary,
musical and even philosophical affinities, hardly a trace of these
filtered through to his business creed, let alone his business
behaviour. Macdiarmid could not project himself to wider audi-
ences. He did not seek publicity, struck most people as reserved
and withdrawn, and performed the social duties of chairman
without enjoyment. But the absence of a public dimension was
also attitudinal, even conceptual. Macdiarmid's lack of direct
experience of production and working people; the narrowness of
his training as an accountant and company secretary; the long
isolation that that implied from the wider currents of public
affairs; a distrust of emotionalism; a single-minded concentration

on the effort of running the firm; an apparently complete conceptual identification between public duty and Stewarts and Lloyds' commercial interests: all of these characteristics militated against social breadth.

It is easy to see, then, how Allan Macdiarmid personified to a high degree the pursuit of growth and efficiency but not social action. The continued expansion of Stewarts and Lloyds is the underlying theme of his writings, statements and actions, the pursuit of 'efficiency', sometimes of an 'economy' which is, however, to be consonant with efficiency, forms a constant and quite explicit refrain. Moreover, growth and efficiency are to be pursued in a balanced and interdependent manner. Growth is never to get out of hand, its pace is to be constrained with a prudent eye on financial, managerial and market limits, it is to be carefully planned. The very rationale of growth is efficiency: expansion is essential precisely in order to secure the economies of scale. At the same time efficiency in the allocation of existing resources is necessary and desirable in itself even when, or particularly when, growth is delayed or inexpedient; it is an imperative if the advantages of growth are to be exploited. The ideal is to carry both growth and efficiency to the highest possible pitch. Although Macdiarmid never formulated an objective of long-run profit maximisation, his concepts are strongly consistent with it and certainly with the idea of an optimising trade-off between growth and efficiency. For Macdiarmid this pursuit sums up the supreme responsibilities of business in the public and national interest. Other priorities are not excluded, for example good public relations, harmonious labour relations or co-operation with government and public bodies, but these are operational necessities along the way rather than primary tasks. Once again, although Macdiarmid did not use the formal concept of 'constraints', his attitude to social action approximates quite well to it.

When Macdiarmid took over Stewarts and Lloyds faced a combination of difficulties: the effects of the General Strike, increased competition at home and in export markets, tariffs in the previously lucrative Australian market, and rapid technological change as many pipe users, for example in the fast-growing oil industry, increasingly switched to weldless products. These suggested massive capital spending on re-equipment just when finance was constrained. With characteristic clarity Macdiarmid put the main alternatives to the board in July 1926:

The one line of policy is to say that in the present state of the country and the Iron and Steel trades, in the present state of our Balance

Sheet, and in the present state of our prestige with the investing public, we cannot contemplate raising additional Capital or Stock and we must, therefore, carry on: making the best of the plant and equipment we have, in the hope that through time and by conservation of profits we shall be able to pay off our Overdraft and restore the business to its old position.

The alternative line of policy is to say that we must definitely make up our minds what additional developments and plant and equipment are necessary to put us abreast of our competitors here and in other parts of the world and that we must find the finance for this purpose.

A great deal could be said on both sides of the question.

For the first outlined line of policy there is this to be said that perhaps the picture is not as black as I have painted it. We appeared to be beginning to make fairly reasonable profits all round just before the Strike. Hickmans and North Lincoln were beginning to produce profits instead of losses. Our position in the Oil business which affects the prosperity of Clydesdale, Imperial and Calder to such an important extent and even Coombs Wood and Phoenix in a minor degree has developed in a remarkable way ...

It may, therefore, be argued that after a settlement of the Strike we shall be able to make reasonable profits sufficient to give a reasonable return on existing Capital.

Even so, I am afraid it would be a long time before we were able to pay off our Overdraft; more especially as on a revival of trade the necessary increase in working capital used will involve an increase in the Overdraft.

Further, to carry out this policy, it would be necessary to bar all Capital Expenditure and if that is done, we may, in a few years, find ourselves in what is ... an essentially unsound position because our plant has become non-competitive and we are faced with enormous Capital Expenditure schemes to bring ourselves up to tone again.

More especially does it appear likely that this would be the case when one considers the enormous wealth of America at the present moment, and the amount of money she is spending on Weldless plant and on research into methods of tube-making, and when one sees how French Tube Makers are developing and how German Tube Makers are systematically combining to cheapen production and eliminate old-fashioned plant.

There is a third line of policy which I have not mentioned. That is to cut some of our losses by shutting down certain plants and withdrawing from trades that have so far proved unprofitable.

I think the answer to that proposal would be that the new plants have not yet had an opportunity of a fair trial given to them.

Further, if that were done I think we should have to face the position that there was little prospect of the remaining plants earning a return on the total Capital of the Company; and the consequences arising from that conclusion.[5]

These early policy papers reflect some of the main keynotes of

Macdiarmid's approach. Large-scale expansion was essential in order to concentrate the production of both tubes and the requisite iron and steel supplies. This in turn was needed in order to keep pace with competitors in terms of both scale economies and up-to-date techniques. Such plans required openness of mind on the question of location as between Scotland and England, and also between existing works and an entirely new one. Capital spending on piecemeal modifications meantime was to be avoided. Efficiency was to be pursued throughout the offices and works, in sales and clerical organisation as well as in production. Moreover, if these pursuits were to be effective several things were also necessary: 'financial balance' and an avoidance of over-capitalisation; avoidance of debt largely, it seems, to re-assure the capital market rather than because of any aversion to borrowing as such; a trimly-shaped board equipped for firm and speedy decision-making; arrangements with competitors to avoid duplication; a practice of ensuring that fallbacks were available if the best outcomes failed. Two more controversial implications of the programme which Macdiarmid did not mention were greater centralisation of decision-making and considerable redundancies. Over and over again he returned to the need for an overall corporate concept or what would now be described as a model. As he put it graphically in one paper: 'An investigation into what appears to be the simple problem of providing plant for making Weldless Tubes above 10″ diameter involves a study of the whole policy of Stewarts and Lloyds throughout all its ramifications... The main problems which face [it] are one and indivisible and cannot be considered and resolved separately or one by one.'[6]

## A polite coup d'état

In 1926, the year of Allan Macdiarmid's final assumption of power in Stewarts and Lloyds, the USC's condition was worsening. The combination of external pressures and internal weaknesses described in Chapter 4 was pushing the firm rapidly towards a cliff edge. Net profits in 1925–6 were a pathetic £80,000. This was bad enough in view of the fact that earnings were probably overstated, depreciation provisions nil, reserves depleted and dividends still non-existent. But worse was still to come. Through 1926 the trend deteriorated, partly because of the General Strike, nor was there any real recovery in 1927. Despite intermittent signs of respite, by autumn 1927 the USC had reached a state of barely disguised crisis. Bank borrowing was over £2¼ million; indebtedness of all kinds had reached the staggering total of £6.7 million; relationships with the National Provincial Bank were more sensitive than ever. In September 1927 the finance committee reported the 1926–7 results to the board and they made grim reading. Only Fox and Rother Vale had made meagre profits, the West Cumberland results were nearly as bad as before, SPT and even Frodingham had turned in losses. Overall, there was a net loss of £40,000, once again, if anything, a serious understatement. Barring some miracle, the future looked bleak. Tax liabilities amounted to over £1 million and the heavy indebtedness meant that annual fixed interest commitments were nearly £464,000. Just to meet these commitments and keep its head above water the USC needed annual profits of about £600,000. Taking a long-term average of past experience such profits were not an unreasonable expectation but in the autumn of 1927 they seemed extremely unlikely. The prospect of an inability to meet current commitments increasingly loomed. So did the fear that the company's creditors might take possession and that a humiliating liquidation and reconstruction might be enforced.[7]

What could the USC board do? The pros and cons of amalgamation were debated yet again. But Walter Benton Jones pointed out in detail the formidable problems of agreeing comparative asset valuations and the earnings prospects of the firms to be merged. One alternative, it seemed, was to sit tight and wait for better times. The chairman, Albert Peech, continued to snatch at any sign of respite, however faint, whilst the operating head of Frodingham, James Henderson, now a director, voiced a typical view with the insistence that 'a marked revival of world demand' was inevitable sooner or later. For H. G. Guedalla 'other concerns are probably in a worse condition ... although it is not much consolation in the circumstances'. Guedalla was against a merger: 'It is probably better to preserve the position as it is.' Effectively, his only remedy was to hold on, to hope for safeguarding tariffs and, somewhat later, to seek 'a financial investigation' from the outside. The relevant views of Scott Smith and Ellis at this time are not recorded: probably they, too, favoured soldiering on but without any real readiness for a drastic internal rationalisation.[8]

It was at this point that the venerable Maximilian Mannaberg planted a bomb. In October he threatened to resign from the board: 'I found for some considerable time that I could do no good by remaining a director of the USC'. There was indeed nothing new about Mannaberg's position as a malcontent. As far back as 1923–4 he had raised the alarm with the Jones's. In characteristically trenchant terms he had then criticised both financial 'shufflings' and discussions about mergers, advocating vigorous internal rationalisation:

> US Company position as we know it – some still don't want to know this – has arisen by an unwarranted and reckless optimism, which always expected the Best and never provided or even thought of the Worst ... the Balance Sheets of the US Co. are as we all know unreal ... My only positive, immediate and *permanent* remedy is, to put your House in order, concentrate all your energy on your own Works, and manage them efficiently.

Now, after a further three years of watching impatiently from the sidelines, Mannaberg boiled over. His view was that 'The United Steel Company is a heterogeneous conglomerate and its name an irony'. Mannaberg's detailed criticisms, as noted by Sir Frederick Jones, are worth quoting:

> Has been dissatisfied with methods at B. Mtgs. from the start. Business thrown at Bd., never properly discussed, says quite unnecessary for AOP [the chairman] to spend so much time in London ... AOP exercises no control over management of branches, has no grasp of the Co. as a whole, does not know what's wrong ... Management at Ickles, Stocksbridge and Workington not gd. enough ... No co-ordination ... Management at works has gone on just as before the amalgamation – nothing has been done to get the best out of them ... Should appt. outsider as M.D. Man with wide vision, imagination and force with free hand to make any changes he considers necessary. Says if not adopted will be forced from outside ... Considers that concern is sounder than Bolckows and if properly mgd. should do well.[9]

What followed was a brilliant exercise in constructive boardroom blackmail. The few colleagues in whom Mannaberg confided reacted with dismay, terrified of the reverberations of a walkout by so authoritative a figure. As Sir Frederick Jones put it, 'His resignation must not be allowed to materialise if it can be avoided ... If it gets to Lewis [General Manager of the National Provincial Bank] it will disturb him and the danger signal will go up at once.' The terms of Mannaberg's ultimatum were conveyed to Sir Frederick. They included not only the appointment of a managing director but also the demotion of Albert Peech from the chairmanship, his replacement by Walter Benton Jones, and the resignation of three other directors, J. E. Peech, Scott Smith and Barber. All this to be combined with a financial reconstruction, 'appreciable cost reduction' and the raising of new capital for investment ('during the last seven years the Branches have been starved').

Sir Frederick was obviously embarrassed by the idea of a concer-
ted move to appoint his own son in place of Albert Peech; but
desperately worried about the financial position and about Mannaberg's threat, probably agreeing with him about the need for
major reforms, and perhaps secretly gratified by the plan to elevate
his son, he went along with the main elements of Mannaberg's
plan. He insisted, though, that Peech should be treated gently.
Guedalla, when consulted, was reluctant to replace Peech and
suggested an independent financial investigation: 'In times like
this we should all hold together and if there is a clear case for
change everyone will surely agree.' To which Mannaberg retorted:
'I know more about the concern and its organisation than they [the
accountants suggested by Guedalla] can learn in say two or three
months time and ... I cannot see my way to await their "recommendation".' Peech himself, obviously aggrieved, denied several of
Mannaberg's charges in a letter to Sir Frederick Jones but was
remarkably non-belligerent: 'Please do not think I am writing in
any fighting spirit.' Peech's ambivalence about his position has
already been noted (see Chapter 4).

Three additional factors helped Mannaberg's plan to go through.
A further worsening of the USC's financial position, evident by the
end of November, strengthened his hand; he was able to produce a
promising candidate for the managing directorship, Robert Hilton,
managing director of Metropolitan Vickers, who impressed the
Joneses when introduced to them privately; and Sir Alfred Lewis,
General Manager of the National Provincial Bank, also impressed
by Hilton, agreed to the changes. Lewis's only proviso, that the
auditors should investigate the USC's long-term debt obligations,
was easy to accept. Some of the harder edges of Mannaberg's
ultimatum were softened. There was no more talk of forcing Scott
Smith or J. E. Peech to resign. Albert Peech was promised a life
pension, made president of the company and consulted about the
press notice so it would be 'in a form which would let him down as
lightly as possible'. The importance of Peech's ongoing sales and
trade representation responsibilities was stressed: even Mannaberg had conceded that 'he is very popular and efficient in this
work'. By the New Year Walter Benton Jones was in place as
chairman and Robert Hilton as managing director. The *coup d'état*
had been remarkably speedy and polite. Its main architect, Maximilian Mannaberg, had cause for self-congratulation. It was to be
the last of his many services to the iron and steel industry but one
which necessarily remained secret.[10]

What were the main strengths and limitations of the two men
whose influence on the USC was to be critically important up to
1939?

The new chairman, Walter Benton Jones (1880–1967) had grown up within a reasonably established upper-middle-class milieu closely connected with coal mining, his father having done well in that industry (see Chapter 4).[11] He spent most of his childhood in Treeton, Yorkshire, near a colliery and a mining village, although his father later moved to more genteel surroundings, a rambling, historic manor house, Irnham in Lincolnshire. Walter's relationship with his father, a widower, a strong personality, apparently very straitlaced, was obviously highly important. The two sons, Walter and Charles, polarised in their responses to him. Walter became the 'good boy', Charles the deviant rebel. After a possibly unhappy period at Repton, Walter read history at Trinity College, Cambridge, and travelled briefly abroad before joining Rother Vale Collieries (RVC) in 1902. He married young, a girl from an established Sheffield business family, and was utterly devoted to his wife until her early death in 1938 after many years of disabling illness. His formation in the South Yorkshire coal mining industry was long and thorough, starting with a period down a mine and training as a mining engineer, the description he was most fond of applying to himself in later years. He became Secretary of RVC in 1906, a director in 1909 and managing director in 1915. After the 1918 merger he soon became executive director in charge of colliery operations throughout the USC.

Benton Jones's conceptual grasp, aura of leadership and practical resilience were all substantially greater than his predecessor's. On the other hand, he was a very different kind of businessman from, say, Allan Macdiarmid. He was cautious rather than adventurous. He lacked Macdiarmid's intellectual brilliance, incisiveness and thrust. He could be laborious in analysis, prolix in presentation, slow in arriving at conclusions, sometimes inclined to ride off at a tangent. His appetite for facts and figures could be excessive. A doughty provincialism, a reclusive home life and a shyness about extra-USC contacts and friendships, were drawbacks in some ways, even though they encouraged a great dedication to hard work.

What, then, were Benton Jones's main strengths? First and most obviously, he cultivated and carried off well what might be called the monarchical aspects of chairmanship. A slight, dapper, bow-tied figure, he was dignified, impressive on public occasions despite his personal shyness, adept at conveying an interest in people, a credible embodiment of devotion to duty, equity and a collective ideal. Not for nothing was he a great admirer of the Royal Family.

Second, his mining industry background had deepened his

character and made him a realist. It differentiated him from other steelmasters, providing a safeguard against complacency, a tougher framework against which to judge the industry's problems. Third, Benton Jones's training as an engineer and his long experience in coal were not the only factors making for a strong concern with efficiency. Also relevant were temperamental drives towards thoroughness, cleanliness and tidiness, the long tutelage of his father, and the influence of Maximilian Mannaberg.

A further important characteristic in Benton Jones, his humanitarian social idealism, probably had more complex origins. Among other things it doubtless reflected his high-minded upbringing, a middle-of-the-road Anglicanism, an awareness of working-class roots not so far back in his own family, a slight unease, perhaps, about his inherited role. Another important influence came from his early and continued links with mining communities. Although such factors produced no break with conventional conservatism, they contributed to a marked social conscience. Not only sexual permissiveness but also *nouveau riche* acquisitiveness, aping the aristocracy, financial manoeuvrings, not earning a living – these things were firmly disliked. The people to be admired were doers of 'good works', nurses, philanthropists, those bent on improving the status and conditions of labour. Benton Jones enjoyed speech-making to workers' representatives and contact with people on the shop-floor whose liking he earnestly sought. He detested unemployment and subscribed to concepts of wealth-as-stewardship, job fulfilment, the factory as a training ground for social virtue and reciprocity, industry as a school for community values.

This mixture of attributes fitted Walter Benton Jones for some of the exigencies of corporate leadership at this time. If his papers to the board demonstrate thoroughness, good sense and moderation rather than superlative generalship or strategic brilliance, they also performed a useful Socratic questioning role and set a tone of high-minded purpose. Here was a man who feared aggressiveness and sometimes laboured the obvious but he could also enlarge boardroom debate and provide many with an inspiriting sense that fundamental purposes were at stake in business life. Benton Jones nicely approximated Bagehot's recipe for a good constitutional monarch: well equipped 'to encourage, to warn and to inspire'. He needed, as it were, a good prime minister and fortunately possessed one. However, this monarchical character was almost purely domestic. Outside the USC, in the wider world of competition and cartels, industrial diplomacy and political influence, Benton Jones's capacities were less marked.

Benton Jones's partner, Robert Hilton, the new managing

director, was in many ways more formidable.[12] Unfortunately, less is known about his background. Like Benton Jones, he had the advantage of wide experience outside iron and steel, although some considered this a drawback at the start. His army experience had doubtless reinforced a temperamental brusqueness, incisiveness and zest for command. Hilton came fresh from a key job as managing director of the electrical engineering firm, Metropolitan-Vickers whence he derived valuable strengths: wide contacts, experience of a new, innovative industry different from iron and steel yet closely linked with it, an understanding of modern management techniques, and, not least, successful practice in the subtle arts of general management.

Hilton's reputation has suffered for several reasons. His managing directorship of the USC ended in 1939 and he did not survive the Second World War to recount his tale or influence a new generation of managers. His working methods, still more his withdrawn temperament, kept him from the public view and made him mysterious even inside the USC. It was Benton Jones who mostly got the limelight. To all of which must be added a certain gracelessness and the fact that Hilton made a number of well-placed enemies inside the firm, people who later implied that he was ruthless and dictatorial, although a measure of these qualities was arguably useful at the time. Yet in any fair assessment of the interwar period in iron and steel this man must be given a central place. Few contributed so much to the advance of professional management in the industry.

True, Hilton's limitations stand out sharply enough: for example, a certain lack of imagination, a direct approach to outside negotiations which could be ham-fisted, an impatience with the City side of financial affairs, occasional naïveté about the industry's convoluted politics, a bluntness which could become merely rude. The man's softer side remained hidden, for instance his enjoyment of 'a yarn at the club', his shy dread of photographers, his aesthetic interests which led him, amidst all the hurly-burly of decision-making, into keen discussions of colour schemes, décor for a new laboratory, tapestries for the directors' room. Even Hilton's constructive work in employee welfare and industrial relations, where he was enthusiastic and committed, remained in the background. His brusqueness had some internal uses and much of his ruthlessness, although exaggerated by some subsequent lore, was, perhaps, necessary. His concepts of management, pithily expressed, emphasised corporate loyalty, professional competence and understanding of men: 'Everybody connected with the side should put the welfare of the Company in front of their personal aspirations'; 'a more scientific frame of mind in place of the rather "rough and ready" methods of the past'; the injunction 'to study and understand the point of view of labour' as a 'first adjustment' for would-be managers. With this corporate emphasis went a rationaliser's enthusiasm for industry-wide reorganisation and for a strong central body. Hilton's contributions to policy-making, management development and control systems were to be enormous. Yet his greatest strengths, perhaps, were not a master strategist's, but were more those of a policy implementer and commander. Written orders expressed clearly with accompanying reasons, relentless techniques of questioning, the ability to move rapidly from one aspect of the business to another whilst retaining a clear grasp of the whole, and flexibility, speed and firmness in reaching operational decisions – at all of this he excelled.

It is necessary to understand how the respective strengths and weaknesses of these two men, Benton Jones and Hilton, interrelated. To some extent their temperamental differences made for difficulties. To a considerable extent they complemented each other: it was basically the king/prime minister syndrome. But for our purposes perhaps the main point is that their attributes combined to produce three dominant biases: a limited propensity for growth, and a marked pursuit of both efficiency and social action.

The circumscribed inclination for growth reflected, *inter alia*, Benton Jones's formative experience in the relatively static coal industry, his natural caution, and his dislike of, and unfitness for, aggressive policies. It also reflected the essential moderateness of both men and the fact that neither was particularly good at the wheeling-and-dealing for mergers, the negotiating manoeuvres with other firms, the power ploys in Whitehall and the City at which a man like Macdiarmid excelled. Again, the pursuit of efficiency reflected engineering influences, the industrial experience of both men, Hilton's adherence to the 'rationalisation' cult and his immediate past at Metropolitan-Vickers. It represented the temperamental preferences for method, control and order, the ideas about harmonising and developing human talents cherished by both men, and their joint capacities for working up performance and morale. The roots of both men's interests in social action were equally deep-seated. No explanation here can neglect Benton Jones's experiences in coal mining and his humanitarianism, or Hilton's labour relations ideas and ardent enthusiasm for industry-wide reorganisation, although other corporate influences were also present, for example an inherited moderation towards labour and a pro-federation bias (see Chapters 8 and 9).

## A traumatic finale

Thus both Stewarts and Lloyds and the USC had achieved decisive changes at the top well before the end of the decade, albeit by different routes. New and younger leaders had been installed and were already pursuing determined policies and radical reforms. But for Dorman Long the managerial watershed was to be very different. It was to prove unconscionably delayed and confused.

At this point it is worth quoting at length some of the firm's internal documents of the period 1927–8.[13] These concern relationships with the bank, a crucible which painfully tested corporate resources, organisational effectiveness and managerial nerves. They add colour to many of the points emphasised in Chapter 4. The firm's financial difficulties compelled an extreme dependence on the bank. The weak financial structure took a heavy toll inside the boardroom in terms of humiliations, pretences and anxieties. Managerial stress and exhaustion were made worse because of an ageing and at times failing chief executive, a thin management team, and confusions about responsibility for finance. Some relationship problems can reasonably be inferred, too, from what follows: for example, the pressure to shore up Sir Arthur Dorman's morale even to the extent of hiding disagreeable facts from him, the exasperation caused by his uninspiring son, Charles, the intrusion of their two-edged father–son relationship, the cover-ups, smoothing-overs and rationalisations anxiously engaged in by the non-family administrators, Byrne and Stubbs.

*T. D. H. Stubbs to Colonel F. Byrne, 1 June 1927:*
Sir Arthur asked me to write to you after some conversation with him. I pointed out the difficulty of my task and asked whether I might say that he was in a depressed mood, and he told me I could say what I liked.

The gist of the conversation was that the Iron and Steel trade was going or had gone to the devil; the bottom knocked out of prices; no orders; increasing imports; blast furnaces closing down. The only possible thing for us to do is to close down our Collieries and Blastfurnaces and Steel Plant, buy foreign material and keep our Constructional Yards in full work. Not only to close down our Steel Works, etc., but to abandon them as useless lumber, so as to avoid overhead charges. Having said that ... he said he would like you to see Goodenough [General Manager, Barclays] and find out what he thought about paying the 6% Preference dividend ... You know the arguments for paying the dividend, so if you have read as far as this without committing suicide, you might consider the proposal of an interview with Goodenough.

*Colonel F. Byrne to T. D. H. Stubbs, 5 July 1927*:
Goodenough wants to see me, Thursday, told me on telephone we have exceeded our limit. I shall see Sir Arthur and Arthur tomorrow, but I don't expect they'll have figures, so can you send a note showing current position giving special reasons for increasing or failing to reduce the overdraft. Will you speak to Charles about this and say that I would have written to him but I am not certain whether he is in Middlesbrough. I rather think he is away on a motor tour.

*T. D. H. Stubbs to Colonel F. Byrne, 8 July 1927*:
[Had seen Sir Arthur and Arthur,] also Maurice Bell was present [It would be possible to prepare some sort of weekly statement on expenditure, if this would help the position with the Bank] Arthur, as you probably know, is very seriously considering the curtailment of output.

*T. D. H. Stubbs to Colonel F. Byrne, 13 July 1927*:
[With regard to profits,] who can say?... As Sir Arthur asked me to give my opinion, I think it is fairly safe to prophecy [sic] that there should be a reduction of at least £250,000 before the end of September, and not unlikely that the reduction might easily be £100,000 more.

*Colonel F. Byrne to C. Dorman, 13 July 1927*:
I have heard from your Father this morning that he has been ordered a complete rest. As these notes are of a reassuring nature, I have thought it better to send them to him rather than leave him in the dark. [Enclosed various figures leading to the conclusion:] There is thus a £4,243,000 increase in Assets – probably accounted for by increased value of Work in Progress. [Then he gave production figures: deliveries of finished material a record, increase in Stocks and Trade debts normal,] They must have been overlooked by our Chairman when he undertook to reduce the Overdraft at the rate he promised.

*T. D. H. Stubbs to Colonel F. Byrne, 14 July 1927*:
As Sir Arthur seemed worried and ill, I thought it better to postpone my holiday. As my house is closed, I am having most of my meals at Grey Towers and am trying to keep Sir Arthur's mind off business after midday. I see him in bed in the morning when we discuss business; after lunch we go round the estate and are farmers. In the evening we talk about everything else except business. I am very glad to say that he is a good deal better, more cheerful and altogether more like himself.

*Colonel F. Byrne to Sir Arthur Dorman, 15 July 1927*:
[Encloses memorandum rewritten in the light of letters from Sir Arthur, Charles Dorman, and Wilks:] If you are disinclined to read [it] in detail, the most important paragraph is I think 9.
   Charles criticises the wisdom of opening a separate account for the Bridge Contracts. Personally, I think it would be wise, but I would like to know whether paragraph 12 of my revised Memorandum should stand.

*T. D. H. Stubbs to Colonel F. Byrne, 19 July 1927*:
[With regard to separate Bridge Account Sir Arthur had already said yes] I referred again to him regarding Charles's criticism, but he told me to carry on and get it done.

*Colonel F. Byrne to Sir Arthur Dorman, 19 July 1927*:
[Had a long talk with Goodenough this morning; the upshot was that there was to be no further increase in the present overdraft] Goodenough was very serious [about this] Wants me to have daily information about our a/c so if he ever has occasion to talk to me, I shall know all about it ... Am arranging this with Stubbs [He asked a lot of questions about Sydney Bridge] I offered for him to talk to Ennis [the manager in charge of that contract] He refused ... He was not so much impressed by Works as he was by Bank balances ... Pleasant but very firm ... I am afraid we can no longer count on the elastic treatment we have been receiving during the last three or four months.

*C. Dorman to Colonel F. Byrne, 21 July 1927*:
[If Goodenough wanted to see Byrne on routine, O.K., but if he was dissatisfied with how DL was being run then it would be very important and Dorman as finance director should accompany Byrne. Dorman had discussed this with his father and Stubbs.]

*Colonel F. Byrne to Sir Arthur Dorman, 21 July 1927*:
[Had seen Goodenough this morning. He had laid Byrne's memorandum before his Board. There would be no further bridge contracts without their concurrence. Goodenough wanted McLintock to satisfy himself as to profits since January 1st and DL estimates for the future, etc.]

*Colonel F. Byrne to T. D. H. Stubbs, 21 July 1927*:
[Heard from Sir Arthur this morning that he wants Charles to handle financial dealings with Goodenough.]

*T. D. H. Stubbs to Colonel F. Byrne, 22 July 1927*:
I think all that Sir Arthur wants is for Charles to be rather more closely in touch than he is. Of course he has probably more information than any one of us for he makes it his business to get his information from each of us.

*Colonel F. Byrne to Sir Arthur Dorman, 23 July 1927*:
[Was afraid Dorman had formed a somewhat false impression of Byrne's last meeting with Goodenough, who, though firm, was very friendly. He had said, 'We are together in this thing'. Byrne felt he and his Board were satisfied DL's position was not a bad one. The DL overdraft was well over £2 million in spite of Goodenough's limit, without any drastic step being taken.]

*C. Dorman to Colonel F. Byrne, 25 July 1927*:
Should you at any time wish me to go with you to see Goodenough let me know and I will make a point of doing it, though naturally I should

like as long notice as possible because I am already pretty full up with engagements.

*Colonel F. Byrne to C. Dorman, 26 July 1927*:
[It would be better if Dorman dealt with, or were present at, all interviews with Goodenough, unless very short notice.] in which case I would do what I could. This I understand was your view but if you have altered it, I am at your disposal to do anything possible.

*Report on the Capitalisation of Dorman Long, by W. H. Davies (Executive Accountant), April 1928, addressed to T. D. H. Stubbs, 14 May*:
I do not need to prove to you that depreciation of plant has not been effected to the extent that it should have been ... the Company has shown a greater profit than would have been done if the policy of the firm had been to create a fund out of which plant could be modernised. The assumption is that too much has been paid away in dividends.

[Quotes Sir William McLintock's report – Depreciation of plant should have been £3,886,260 – Actually charged £874,170 – Capital Expenditure to Revenue, £537, 460] My first point ... is that Dorman, Long's capital should be written down by £2½ million.

[Detailed comments on fixed assets or capital expenditure to date of £9,875,000; by-products plant seemed under-valued – Carlton Iron Works appeared to be reasonably valued. On ironworks and coke ovens] some very extraordinary figures [overall suggests writing down by one third of value or about £1.3 million:] It is recognised that the coke ovens are old and the blast furnaces almost entirely out-of-date [Steelworks: Redcar and Acklam completely up-to-date, recent improvements in Britannia, but value of Clarence should be written down by £¼ million.]

[Figures on urgently needed capital expenditure, leading to the conclusion,] £2 million will make Dorman, Long an efficient steel producing unit.

[Recommends complete financial reconstruction, arranged with] a strong finance company ... I would not venture to make such suggestions were there not strong public faith in D.L. and Co.

Something drastic needs to be done.

Neither financial cliff-hanging nor managerial overstretch were to recede for some considerable time yet. As the turn of the decade approached published profits slightly recovered (1928, £353,000; 1929, £391,000; 1930, £394,000) but they could not cover depreciation or urgently needed re-equipment and were barely adequate even to meet prior charges and bank interest. In 1931 they slumped to a derisory £30,000. Plant closures and large-scale lay-offs of employees were resorted to, although there was little rationalisation in production or even in other (financially easier) fields. Few of the managerial defects were remedied. The top management team still basically comprised the two

octogenarians, Sir Arthur and Sir Hugh, the limited and uninspiring sons, and a few elderly and overworked administrators.
And this group was further depleted with the deaths of Charles
Dorman and Colonel Byrne in 1929. Despite Dorman Long's size,
vertical complexity and overseas ramifications, there was hardly
a trace of the control systems employed by a firm like Stewarts
and Lloyds (see Chapter 4).[14]

To reorganise the whole edifice woould have been difficult
enough without the addition of yet another large company with
comparable problems. But this is precisely what now occurred as
a result of an amalgamation with the neighbouring Cleveland
firm of Bolckow Vaughan.

The historic firm of Bolckow Vaughan had recently been
through even worse times than Dorman Long. It was of
medium-to-large size, fairly highly vertically-integrated, with a
finished steel production about half that of Dormans and capitalisation of upwards of £8 million. Swingeing losses in the early
1920s reflected not only the depression but also the non-recovery
of large sums of money from the government, over-priced acquisitions, a scale of debenture borrowing probably even more misjudged than Dorman Long's, and general mismanagement. In
1924 a shareholders' revolt, followed by a ponderous but critical
report by outside accountants, led to the chairman, Sir Edward
Johnson-Ferguson, being forced to resign. The appointment of a
managing director, recommended by the accountants, did not
take place until 1926. The improvements introduced by this man,
Sir Holberry Mensforth, had not yet borne fruit and the firm was
still extremely weak.[15] In view of the geographical proximities
and longstanding links between Dorman Long and Bolckow
Vaughan, the idea of a merger was hardly new. In spring 1927
negotiations started, apparently on the initiative of Roland
Kitson, Bolckow's chairman.

The scope for horizontal rationalisation between the two firms'
production and constructional interests was one obvious motiva-
tion. Another and probably bigger one was both firms' financial
weakness and their dependance on bankers who were keen on a
merger and capable of exercising leverage to bring it about. If
anything, at this time, the fashionable cult of large-scale mergers
in the cause of 'rationalisation', much discussed in Whitehall and
the City, excited banking circles more than it did steel industry
chieftains. No less a body than the Bank of England, now on the
brink of rationalising interventions in iron and steel (see Chapter
6), became interested in the Dorman Long, Bolckow Vaughan
merger. It soon became clear to the beleagured boards of both
companies that if they wanted substantial financial assistance,

they had better get together quickly. Not only that: such a bilateral grouping was intended to be a first step towards an even more ambitious trilateral one. The longer-term goal was to be a comprehensive North East Coast amalgamation scheme also involving South Durham and Cargo Fleet. With this end in view, not only would immediate support be available from the two companies' bankers but the Bank of England itself would assist in raising more money later for the larger scheme.[16]

There is no sign that the managerial implications of all this were considered. Enormous assumptions about attainable rationalisation economies, corporate assimilability and managerial capacities were largely implicit. The financial institutions were constrained by their traditions of non-intervention. The thought of dislodging so distinguished a figure as Sir Arthur Dorman and installing a new top management to make a success of the Dorman Long, Bolckow Vaughan grouping and to prepare for the even larger North East Coast scheme would have gone against the grain. There is no record of Sir Arthur's own reactions. Probably he was tired and resigned to the change. Certainly he was prepared to defend it publicly; no doubt a spark of his old fighting spirit remained. After all, the merger represented something of a replay of his characteristic growthmanship, a reminder of great past adventures: not only a source of financial rescue but also an apparent breakaway from deadlock and, with the absorption of the historic Bolckow firm, a salve to his long-wounded pride.

In fact, the final stages of the old man's régime were approaching at last. The following year brought further economic and financial tribulations, also difficulties consequent on the merger. Then, in February 1931, Sir Arthur Dorman died. The obsequies at St Mary's Church, Nunthorpe, including a lengthy procession of Dorman Long employees, were a suitable marker of the end of six decades of business leadership, of the passing of an 84-year-old industrial monarch who had become an almost legendary local figure, a symbol of patriarchal continuity with the Victorian age. Within twenty-four hours of the funeral the Dorman Long board proceeded to pay a remarkable tribute to both longevity and tradition. In a move that was greeted with incredulity in some parts of the firm they appointed Sir Hugh Bell, the dead man's 87-year-old colleague, as his successor. But Dorman Long's long-delayed managerial changes were fast approaching. For by early July Sir Hugh also was dead. His successor as chairman was to be Charles Mitchell, a non-family man who had been ascending rapidly within the firm's top echelons.[17]

Mitchell's appointment represented a shift away from dynastic and local Middlesbrough influences, an assertion of the idea that Dorman Long required not only a new broom but also wider skills and contacts. A man in his mid-fifties, Mitchell's earlier career had been in constructional engineering and the supervision of large contracts at home and overseas. Entering Dorman Long in 1924 as manager of the bridges department, he soon became Manager of the London office and yard. Because constructional work was the most expansive and publicly prestigious part of the firm (see Chapter 4) his reputation grew. Since the London office did much work relating to overseas activities, trade investments and banking arrangements, his grasp of large parts of the business increased quickly. All this, combined with the signal lack of top management succession material elsewhere in the firm, made for a rapid ascent. By 1928 we find Mitchell not merely reporting to the board about railway schemes in Colombia but also making important operating decisions in South Africa and writing to Sir Arthur Dorman about a discussion he had had with Goodenough of Barclays. When Colonel Byrne died in 1929 Mitchell must have seemed the obvious choice to succeed him on the board.[18]

Thenceforward Mitchell's career accelerated. He played an active role in the culminating stages of the merger negotiations with Bolckow Vaughan and in the initial phases of the merger discussions with South Durham and Cargo Fleet (see Chapter 6). Still based in London, trim, brisk and competent, a contrast to the two octogenarians and their unimpressive heirs, Mitchell was a success in financial establishment circles. He seems to have impressed Goodenough. In early 1930 he was offered a position in the Bank of England's newly-formed Bankers' Industrial Development Company (BID Co). It was probably this which helped his next big move inside Dorman Long, promotion to a joint managing directorship, alongside Arthur Dorman, at a salary of £10,000. Within the Bank of England Mitchell was reported to have complained about the 'old deadheads' in charge of the firm and to have suggested that 'someone in the City should buy up the shares and appoint a first class man (such as himself) to carry the remainder of the shareholders and join up with South Durham'. Montagu Norman and Sir Andrew Duncan 'both thought if anyone could clean up the situation Mitchell could'.[19]

Much now depended on how real Mitchell's new power was and how he would exercise it. The first indications were mixed. The old dynastic element was pushed aside but not dislodged. Arthur Dorman became vice-chairman 'with the duty of undertaking the external representation of the Company upon Associations and Institutions, national and international', a kind of House of Lords position analogous to Albert Peech's in the USC (see p. 98). Maurice Bell and Walter Johnson were relieved of their executive responsibilities for the collieries and the works respectively but remained on the board. These rearrangements probably left sore feelings in still-influential quarters. In their place Mitchell proposed to run the firm as an executive chairman, largely through a non-directorial management committee composed of the existing departmental heads. But he himself was to stay in London. His capable co-director, Laurence Ennis, still in charge of the vital Sydney Bridge contract, was not brought back from Australia (he was later to become managing director). The records suggest no searching examination of management structures and personnel, no early retirements or reallocations and, perhaps more significant, no significant new outside appointments. In fact, the first signs were that Mitchell's 'clean-up' would be somewhat half-baked, though whether he had the leverage to do much more is questionable. He possessed neither a formal surgeon's mandate to wield the knife, nor the prestige of a 'new broom' outsider like Robert Hilton, nor even much of a personal power base inside the firm. More, whether he had the actual abilities to press reorganisation home is uncertain. After all, his expertise had been in bridges, selling and external diplomacy, and he lacked experience of general managership either in iron and steel or elsewhere. Also, to be fair, it would have required a species of genius to perform all that was expected of him.[20]

Thus Dorman Long's managerial watershed was less decisive as well as later than either Stewarts and Lloyds' or the USC's. An ambiguous outcome both reflected and contributed to an increasing trend for the company to lose much of the control of its own destiny. It was now set upon a path, partly determined by others, in which there was little room for manouevre. The hope was that Mitchell would clean out the firm's Augean stables, make a success of the Bolckow Vaughan merger and implement the vision of a grander North East Coast Union. A more herculean version of the efficiency-plus-growth ideal can hardly have been sought.

## Conclusions

To any student of business policy certain features of these various changeovers stand out clearly. One obvious theme is the difficulty of replacing weak, inadequate or elderly top managements. This did not merely reflect the passivity of bankers and outside shareholders. Nor was it even just a matter of the established leaders' cussedness and obstructiveness, or their colleagues' sloppiness. In the former case concepts of duty were also at work, in the latter attitudes of loyalty and friendship which could play a significant and more positive role in other ways (see Chapter 4). Another obvious point concerns the vital importance of management succession material: well-planned in Stewarts and Lloyds, partially available, largely through luck, in the USC, sadly lacking in Dorman Long. A significant collective aspect of these changeovers is their effect on the dynastic factor. Although family influences and inheritance remained strong in these iron and steel companies, as in many other British industries, they were now rather more qualified by ideas of professional competence.

However, it is over the hypotheses concerning managerial specialisation and corporate phases that the main questions occur. Were external economic influences limited, as the hypotheses suggest? Did the characteristics of the new chief executives concentrate around partial mixtures of growth, efficiency and social action as predicted?

With regard to the first question, the restricted influence of economic fluctuations on the changeovers is surely clear. Capital market disciplines were weak. Even top managements ill-attuned to the new circumstances possessed a remarkable ability to ride out the depression. If economic imperatives required new leaders, it took Stewarts and Lloyds five years to bring this about, the USC eight years, and Dorman Long eleven years (even longer if the subsequent limitations of the Mitchell régime are reckoned with). Even when the changeovers occurred, they represented no complete, sensitive adjustment to current or expected economic conditions. Despite the gloom of 1925–6, Allan Macdiarmid's succession presaged a stronger emphasis on growth as well as efficiency. Although Hilton and Benton Jones's assumption of power in early 1928 signalled a retrenchment which was timely economically, this was to be strikingly qualified by social action, particularly in terms of employment maintenance. If Mitchell's promotion in 1931 largely represented the external bankers' economic ideal, this was a misjudgement of both personality and corporate factors. Just as the timing of the changeovers mainly reflected death, chance and boardroom politics, so the orientations of the new men largely transcended current and expected economic trends.

In examining whether these orientations specialised around dominant tendencies, as predicted, I have drawn on data relating to childhood influences, professional training, industries worked in, as well as temperament, ideals and abilities. As a result a tentative typology emerges. Allan Macdiarmid appears to personify a combined pursuit of growth and efficiency, with social action lagging behind. Despite their many differences, Walter Benton Jones and Robert Hilton appear to exhibit marked combined tendencies towards efficiency and social action, with growth coming a poor third. Charles Mitchell, a more shadowy figure, poses the problem of ideas and aims versus capacities proportionate to the task. In his case the evidence suggests growth and efficiency as strong desiderata, insufficient power and ability to back them up, and a consequent possible hiatus for both. But these tentative judgements require fuller investigation in the next few chapters. So does the hypothesis that managerial specialisations provided the lifeblood for further phases of development of the firms.

### Notes

1  S/BM, 27.3.24, 6.3.25. The 'caretaker' hypothesis was suggested to me by Wilson's grandson, W. N. Menzies-Wilson.
2  S/CB. S/AGM, 27.3.26. S/BM, especially 27.3.25, 25.9.25, 20.11.25, 11.3.26, 10.3.27, S/GPC, 4.6.25, 16.7.25, 20.8.25, 24.9.25, 17.12.25, 28.1.26. F. Scopes, *The Development of Corby Works*, p. 40–2.
3  Scopes's *Development of Corby Works*, no mere celebratory panegyric, is a useful summary which includes much information on Stewarts and Lloyds' raw material activities and some important policy papers, particularly of Macdiarmid. The latter emerges as hero quietly rather than explicitly. For Burn, Macdiarmid was a man of 'daring and foresight', 'a dissident progressive', 'the outstanding personality of the industry'. See D. L. Burn, *The Steel Industry 1939–59* (1961), pp. 57–60, 67–8. See also D. L. Burn, *The Economic History of Steelmaking 1870–1939* (1940), p. 57.
4  The following account is based mainly on archival materials but also on interviews with ex-senior officials, some of whom knew Macdiarmid personally, and with members of his family. The following sources add little: *The Times*, obituary, 15.8.45; *Allan Campbell Macdiarmid* (monograph: privately printed, 1946), and *Who Was Who*.
5  S/BM, 8.5.25, 29.1.26, 26.3.26, 15.7.26, 8.8.26, 21.10.26, 18.11.26, 16.12.26, 20.1.27, 14.7.27, 18.8.27, 15.9.27, 19.10.27, 17.11.27. S/GPC, 24.9.25, 10.3.26. Scopes, op. cit., pp. 50–4, 164–7.
6  S/BM, 14.7.27, 18.8.27, 19.10.27, 17.11.27. S/GPC, 18.10.27, 24.10.29. S/CB. S/AGM, 30.3.28. Scopes, op. cit., pp. 50–4, 168–171.
7  U/DM, 12.1.26, 9.2.26, 9.3.26, 13.7.26, 10.8.26, 14.9.26, 12.10.26, 11.1.27, 8.2.27, 8.3.27, 10.5.27, 14.6.27, 9.8.27, 6.9.27. General Matter File (006/10/3), Sir F. Jones to C. V. Ellis, 24.7.26, Sir F. Jones to Walter Benton Jones, 27.7.26, W. Benton Jones to Sir F. Jones, 28.7.26, Sir F. Jones to M. Mannaberg, 12.8.26, Notes, January 1927. U/FJ. Amalgamation 1927 (006/10/3), Chief Accountant to Sir F. Jones, 11.2.27, memorandum by Chief Accountant, August 1927, Memorandum by W. Benton Jones, 10.5.27.
8  Amalgamation 1927 (006/10/3), Report by W. Benton Jones, August 1927. U/DM, 8.11.27, 6.12.27. General Matter file (006/10/3), J. Henderson to Sir F. Jones, 10.8.27. U/FJ, Notes by H. G. Guedalla, August 1927. U/RA, H. G. Guedalla to Sir F. Jones, 28.11.27.
9  U/RA. Amalgamation with various concerns, 1921–3 (006/10/3), M. Manna-

berg to Sir F. Jones, 4.3.23. U/FC/C, M. Mannaberg to W. Benton Jones, 6.12.24 and to Sir F. Jones, 14.8.24.

10 For the last three paragraphs see particularly U/RA. Also U/DM, special meeting, 8.12.27 and U/FJ, especially Sir F. Jones to Sir W. Peat, 15.2.28.

11 The following four paragraphs are based partly on archival evidence, partly on interviews. Some useful points also emerge from Andrews and Brunner, interviews, 1950.

12 The following three paragraphs rely on the archival evidence more than in Benton Jones's case, mainly reflecting the survival of Hilton's voluminous letter books and the fewer survivors who knew him. But again, the author's interviews with retired senior officials of the USC contributed.

13 All of these are drawn from D/FF.

14 D/DM. D/FC. D/FF. D/AGMs. D/CI.

15 Historical material on Bolckow Vaughan (210(d)/20/50). North East Coast Steelmakers (210(d)15/23). Bolckow Vaughan, Director's Special Minutes (1066/13/7). Report on Bolckow Vaughan and Company Ltd by Sir W. Peat and Sir W. Plender, August 1924 (1066/25/5). Sir Holberry Mensforth (1871–1951) had been Director General of Factories at the War Office, 1920–6, after long experience in the engineering industry.

16 For earlier merger moves between the two firms see D/DM 11.3.19 and 8.5.19. For Dorman's side in the 1927–8 negotiations see D/DM, for Bolckow Vaughan's see Directors' Special Minutes, op. cit. F.C. Goodenough, General Manager of Barclays, Dorman Long's bankers, was reported as 'strongly in favour of a combination' in D/DM, 5.5.27. For the Bank of England's role see SMT 2/93. There were difficulties between the two companies' boards, particularly about the relative valuation ratio, and fears that the debenture holders would oppose extra bank borrowing, the composition of the new board of directors, and an attempt by Bolckows to get their name included. It is unlikely that these problems would have been resolved if the financial institutions had not used their substantial leverage over both firms in favour of the merger. For the final terms see Letter to Debenture Holders and Shareholders of both companies (05854) and D/AGM, 17.12.29. On the estimated merger savings D/DM, 17.7.30 stated that 'useful figures were unlikely before the end of the financial year'. By 13.11.30 these were received but no details were given.

17 D/DM. D/FC. Bridge Department (1066/8/4). Of the important arrivals from Bolckow Vaughan, R. D. Kitson, its chairman and a figure in the City, became chairman of Dorman Long, and C. L. Dalziel contributed financial knowledge to the Dorman Long board. Sir Holberry Mensforth, evidently finding an inadequate outlet in Dorman Long, became chairman of the English Electric Company in autumn 1930 and a year later resigned from the board: D/DM, 23.9.30 and 22.9.31. D/DM, 17.2.31, 6.7.31. *North-Eastern Daily Gazette*, 11 and 12.2.31, 30.6.31. Interviews with retired Dorman Long officials.

18 *Northern Echo*, 29.3.34. D/DM, especially 20.3.28. D/FF. D/CI, particularly report by C. Mitchell and S. W. Rawson on visit to South Africa, 15.1.28, in D/CI, 00593. Bridge Department (1066/8/4). D/AGM, 17.12.29.

19 D/DM, especially 6.5.29, 17.12.29, 18.3.30. D/CI, C. Mitchell to T. D. H. Stubbs, 13.12.29 on a clean-up of a trade investment. Interviews with retired Dorman Long officials. BE: BID 1/79, Sir G. Granet to C. Bruce Gardner, 10.12.30; SMT 9, 16.3.31.

20 D/DM, especially 21.7.31 and 22.8.31. Interviews with retired Dorman Long officials. Laurence Ennis's longstanding seniority in the firm's works management can be glimpsed in Britannia Works (803/1/2), NESCo (1066/13/7) and Chief Chemist's letter books (203/1/77). He joined the board in 1924 but

was heavily involved on the Australian side from at least 1925, see particularly D/CI (1066/5/5). His specialisation and long absence from the centre probably combined with human and public relations limitations to put him out of the running as an alternative to Mitchell.

# 6 Mergers and investment, liaisons and wars

This chapter discusses how the new managerial régimes introduced in the last chapter responded to the challenge of growth in the 1930s, a story which runs the whole gamut from bitter failure through cautious stalemate to triumphant success. Dorman Long failed in its most significant expansion project, the USC achieved moderate growth tinged with some failures, Stewarts and Lloyds bounded forward dramatically. Behind these contrasts, once again, lay variations in the decision-making process and in managerial orientations to growth. How far success or failure in growth was a reciprocal of varying performances in the fields of efficiency and social action will emerge from the following chapters as well as the present one.

## Dorman Long's grand merger débâcle

Let us look first at Dorman Long's most substantial growth challenge from 1929 onwards, the plan for a grand North East Coast merger with South Durham and Cargo Fleet. The negotiations for this merger continued for three and a half years, attracted much public attention and eventually met with crashing failure. For this failure the main responsibility can be attributed to deep-seated, cumulative managerial factors within the firm that was the prime mover, Dorman Long.

The South Durham Steel and Iron Company was a smaller, more successful and more profitable firm than Dorman Long. In 1928 it had achieved a complete fusion with the Cargo Fleet Iron Company, with which it had long had a close relationship. Both groups had the advantage of more recent formative periods (mainly in the 1900s). Mainly situated in West Hartlepool and Stockton-on-Tees, they were relatively free of unintegrated bits-and-pieces of works, and were more compact than Dorman Long. The Furness family, the main initiators and still important shareholders, had been, on the whole, enterprising. Their current head, Viscount Furness, had been chairman since 1917. A more formidable personality, Benjamin Talbot, managing director of Cargo Fleet since 1907 and now operating chief of the whole group, was a man of great drive and shrewdness, an individualist, a leading figure in the industry. Management had been growth-and-efficiency orientated, in some ways analogous to Stewarts and Lloyds. Financial policies had been similarly cautious and, for iron and steel, profits had been both stable and high. By 1929 the combined group was producing slightly over 400,000 tons of rolled steel per annum and had a capitalisation of £2.3 million, including debentures of slightly over £1 million, plus a small bank overdraft. This compared with a Dorman Long equivalent production of 940,000 tons (to which must be added, of course, its substantial constructional and other activities), a capitalisation of over £17 million, including debentures of £6 million, and a bank overdraft of about £2.5 million. The financial disparity was dramatic.[1]

The broad economic arguments for a merger appeared strong. Dorman Long's Britannia works and South Durham's Stockton plate mills could be closed. The production of heavy joists and sections, rails and heavy and light plates in particular, could be concentrated. In addition, there was talk of economies in raw material distribution. Estimates of the potential annual savings depended on differing assumptions about cost–output levels, raw material and other costs, capital spending associated with the merger, etc., and they also fluctuated with changes in economic conditions between 1929 and 1933, varying between £300,000 and over £1 million.[2] But in 1930 an extra political and psychological twist was added. The idea of regional mergers became the centrepiece of discussions about steel industry reorganisation within both Whitehall and the BID Co. As one of two or three obvious 'regionalisation' projects, the North East Coast scheme entered the national limelight. It became a symbol of whether the steel industry was in earnest about 'rationalisation'. The answer to that question might well influence, in turn, first, whether the industry would receive tariff protection and then, later, after tariffs were brought in, whether they would be continued at high levels.[3]

The central problem was how to engineer a union between a large, weak group, which was also over-capitalised and debt-ridden, and a smaller, stronger one. For Dorman Long the idea was attractive in so far as it promised the elimination of an aggressive rival, an access of productive strength and money for desperately-needed modernisation. For South Durham, though, the scheme meant going in with a firm which they saw as unwieldy, inefficient, uncongenial and likely to dominate because of its size. The financial incompatibilities were no less stark. A viable merger would require Dorman Long's capital to be drastically written down, probably by at least one-half, and its debts to be substantially reduced. This would mean large sacrifices for its bankers, debenture holders and preference and ordinary shareholders together with a reconciliation of their sharply conflicting claims. Otherwise, the creditors and shareholders of South Durham would lose out and would be liable to reject the scheme.

Only some remarkable combination of the following conditions could solve the problems: (a) a dramatic improvement in the management, fortunes and prospects of Dorman Long; (b) big concessions to, and forbearance on the part of, South Durham's leaders; (c) substantial outside finance; and (d) super-diplomacy to induce all-round sacrifices from the numerous and disparate groups with financial stakes in both firms.

The main negotiations began in December 1929, in autumn 1930 they petered out due to management upheavals on both sides and political uncertainties, in November 1931 they revived, then continued until summer 1933. An exceptionally varied *dramatis personae* included not only Charles Mitchell, leading for Dorman Long, matched by Benjamin Talbot for South Durham in the first phase and A. N. MacQuistan, his successor as managing director, in the second, but also Charles Bruce Gardner, managing director of the BID Co, a highly active intermediary, and, from time to time, F. C. Goodenough of Barclays, and the accountants, Peat, Marwick, Mitchell. Key onlookers were Montagu Norman and Sir Andrew Duncan inside the BID Co, leading NFISM people and, more spasmodically and remotely, even Sir Horace Wilson at the Treasury and members of the Cabinet. The discussions were tortuous and often fruitless. Much frustration lay behind laconic entries like 'no new developments' (in the Dorman Long Directors' Minutes) and 'many meetings, no progress' (in the BID Co records). Occasionally exasperation broke out as when a South Durham director confided, 'I am sick of these negotiations dragging on as they have done'. Some important developments had to be frozen. Both firms had to be watched in case their investment projects in the meantime unduly favoured one or the other or cut across the merger scheme. In 1930, for example, the BID Co vetoed finance to help Dorman Long modernise its coke ovens, an urgently needed project. There were strains when sharp price competition broke out between the two firms in 1931, and also when dramatic environmental changes through 1931, 1932 and early 1933 upset previous calculations and delicately worked-out compromises.[4]

The question of why the negotiations dragged on so long is largely indistinguishable from the more important issue of why they eventually failed. Analysis can usefully concentrate on the four critical factors for a breakthrough identified above. To begin with, Dorman Long's bargaining stance continued to be weak. Early on they overplayed their hand by blandly assuming an asset ratio based on existing capitalisations and insisting on a straight 'Dorman Long' title for the new combine. They repeatedly underestimated the problems of persuading the South Durham shareholders. Mitchell's negotiating ploys could be ill-judged. The BID Co described a scheme he had for getting rid of Talbot as 'wild-cat', advised him to 'go slow with his pushful personality' and effectively told him another of his schemes was naïve.[5] On the other hand, he showed great assiduity and his flexibility increased. More important setbacks were miserable financial results in 1931 and 1932 and the eventual loss on the Sydney Bridge contract because of currency transfer difficulties. Although these setbacks encouraged Dorman Long to accept the long-resisted writing-down of its capital, they also exacerbated its image of weakness and inefficiency. Mitchell's internal rationalisation continued to be half-baked (see Chapter 7). Ironically, his failure to reorganise vigorously in Middlesbrough partly reflected his concentration on the merger negotiations in London, a striking example of growth pursuits sacrificing efficiency via the opportunity costs of managerial effort, in this case to the extent of damaging even the growth priority itself. The opinion that Dorman Long were negotiating from weakness was widespread, common inside the BID Co and particularly astringent within South Durham: for example, 'The Dorman Long Group was so weak that he was not prepared to be associated with them' (Talbot in May 1930) and 'I am not going to be a party to bolstering up or perpetuating a situation that cannot possibly be run on commercial lines' (MacQuistan in April 1932).[6]

Understandably, there were shortfalls in the South Durham leadership's responses. An early obstacle was the important shareholding interests of the Furness family. Lord Furness was prepared to sell out but it took time for his early insistence on a high price and a full cash settlement to soften. Talbot was a harder nut to crack. He took a tough view and stood out for a key negotiating role to which his abrasive manner was unsuited. From autumn 1931 MacQuistan, South Durham's new general manager, conducted the negotiations, which helped to ease things along. But Talbot's key influence inside South Durham continued. He only agreed to throw his public weight behind the proposals after Dorman Long made big concessions on the capitalisation issue and after his own position was safeguarded. He was to be appointed deputy chairman of the new company, and consulting expert at a salary of £5,000. Given Talbot's abilities this was fitting. But it is likely that he remained privately sceptical; more than likely, too, that he would have made an awkward bedfellow if the merger had succeeded.[7]

On the third precondition for a breakthrough – substantial external finance – the BID Co's attitude was critical. It vetoed the idea of buying out the Furness or other special interests. As Montagu Norman put it with characteristic pithiness: 'This matter must "stew"...it often happened the less keen one appeared to get people in the more anxious they became.' More important was the BID Co's general caution during the difficult economic and capital market conditions of late 1930 and 1931. Even in January 1932 Montagu Norman was still worried about abilities to fulfil 'moral commitments...towards the North East Coast' so that Bruce Gardner had to assure him that, although £1 million was required to make the scheme effective, 'it could be brought off without any money, as by reorganisation the firms can reduce their costs'. There can be little doubt that if the BID Co had been able to engineer specific pledges of substantial outside finance, this would have greatly eased the log-jams of 1930–2, but it was not. A graver contention, perhaps, is that the BID Co did not use its leverage to get the management position inside Dorman Long investigated and improved. Large-scale finance would really only have been justified under such conditions. But it was committed to support Mitchell. In any case, despite a growing industrial expertise, it was deeply attached to an arms' length position and unprepared for an actively interventionist role.[8]

The final and most complex breakthrough condition – the engineering of all-round sacrifices from bankers, creditors and shareholders – was partly a reciprocal of the other three. Unfortunately, the need for secrecy on the detailed terms; the pressure meantime to soften up important institutional shareholders, in private; the consequent risk of an eventual underdog backlash among the many small shareholders involved, some of them influential; the ethical niceties of balancing the various categories of claims; the enormity of the required revaluation of Dorman Long's capital: all this made for inordinate problems.

Dorman Long's bankers, Barclays, played a central role partly because of the £2½ million overdraft, partly because they were the trustees of the firm's 5½ per cent debenture holders whose critically large stake amounted to over £5 million. This dual role was highly sensitive. Key questions became: would Barclays demand even more security for their loan plus the application of merger funds and post-merger profits to reduce the overdraft? But if so, could they simultaneously, *qua* trustees, recommend the debenture holders to accept a scheme in which their (Barclays') actions had reduced the latters' prior rights over the Dorman Long assets? Cross-pressured and embarrassed, Barclays actually did both these things, earning some subsequent legal castigation to the effect that they had used their power unfairly to obtain 'substantial advantages'.[9]

Consequently, it was the Dorman Long debenture holders who were squeezed into the position of appearing to bear the greatest burdens. They were finally asked to exchange half their debentures for preference and preferred ordinary shares and also to accept much less priority over profit for the remaining half. At least some of them were bound to object, legally endangering the scheme, and this they did. In theory, their proposed demotion should have made things easier for the South Durham stakeholders. For them the final terms were indeed advantageous. But here, too, an articulate and influential minority objected. A number of important West Hartlepool interests with stakes in South Durham feared a post-merger transfer of resources from their locality. Probably more important were shareholder fears of financial insecurity inherited from the long depression years, a feeling that South Durham, 'their' company, a successful and profitable one, would be swamped, that the scheme was unwieldy, and that Dorman Long was inefficient and weak. In the event it was the legal actions taken by these two dissenting minorities of stakeholders which finally wrecked the scheme in autumn 1933.[10]

It is tempting to allocate responsibility for this débâcle liberally between Dorman Long, the South Durham leadership, the BID Co, Barclays, and the shareholder revolts, not to mention wider economic and capital market conditions. However, the view taken here is that the main responsibility lay with Dorman Long's management but in a wide and long-term sense. It was the firm's weakness which made it a poor, over-anxious and potentially swamping suitor, which rendered the financial machinations so tortuous, threatened to maltreat the debenture holders or alternative stakeholders in the firm, and made the whole idea suspect to one South Durham interest after another. For this weakness, in turn, as I have argued, Dorman Long's successive leaders were largely to blame. That responsibility partly related to Charles Mitchell but also, more importantly, to the preceding managerial régime of the 1920s or even earlier. The over-expansions of 1916–20, the 1923 debenture issue, the control–efficiency failure and the overall managerial weakness of the 1920s must bear the heaviest share of the blame.

A particularly long managerial obsession with growth and a failure to move to an efficiency phase as a necessary reaction had produced a crisis by 1931. This was redeemable only by a dramatic, quick-acting growth–efficiency combination calling for exceptional managerial powers. By then, anyway, external forces had intervened so that management partially lost control. Merger pursuits, largely pushed onto the firm from outside, escalated. In addition to everything else, these pursuits further deflected time and energy from the actions on efficiency which were now more necessary than ever (see Chapter 7). The result was much débris and confusion in 1933–4 from which a new and more definitive managerial régime emerged – but that is another story.

### Growth triumphant: Stewarts and Lloyds

At the other end of the spectrum from all this lies the story of Stewarts and Lloyds' triumphant expansion, under Allan Macdiarmid's leadership, between 1929 and 1939. By the standards of the firm itself, the steel tube trade, the iron and steel industry and even, in some ways, industry at large, the growth achievement of Stewarts and Lloyds during this period, using 'growth' in the large, strategic sense emphasised earlier (see Chapter 1), was outstanding.

Stewarts and Lloyds' growth involved seven main processes, each showing a consummate exercise of power, judgement and drive: (1) takeovers of tube firms; (2) liaisons with tube firms; (3) leadership in the domestic tube trade; (4) international operations; (5) iron ore acquisitions; (6) plans for Corby; and (7) gaining finance and public support.

(1) *Takeovers of tube firms.* During the period Stewarts and Lloyds spent over £1½ million on the purchase of seven tube manufacturing firms and the part acquisition of two more. The object was not 100 per cent monopoly but rather the elimination of awkward competitors, the concentration of production and a strengthening of the firm's (and Britain's) power position *vis-à-vis* the international trade. Most of the firms were picked up cheaply during the depression phases between 1928 and 1932. It was an advantage that mobile, semi-speculative tendencies in much of the trade made some firms readier to sell out for quick gains than would be the case in heavy industry. A few were harder to crack. Shares were quietly bought in the open market to help in bringing to heel one important, long-established and

proud firm, the Scottish Tubes Company.[11] With British Mannesmann, part German-owned, an exporting, relatively efficient
firm, the largest remaining domestic competitor after 1932 and a
longstanding thorn in Stewarts and Lloyds' flesh, there had been
repeated merger forays since at least 1916. Its purchase in 1935,
jointly with Tube Investments, brought in a further 10 per cent of
the home trade.[12] Most of the production facilities gathered from
these takeovers (though not Mannesmann's) were closed down in
the interests of rationalisation. The process left a residual tail of
some twenty small companies outside Stewarts and Lloyds,
mostly high-cost firms in the Midlands, with no more than 10 per
cent of the home market between them.[13]

(2) *Liaisons with tube firms.* Where Stewarts and Lloyds
lacked the capacity or desire to take over a key firm it sought a
special agreement with it. This was the case with Tube Investments (TI), a holding company group mainly of small units producing precision-made, smaller types of tubes. There was already
some direct competition with TI and considerable potential for
each to enter the other's trade. Earlier moves for a merger, in
1928, had foundered. In 1930, though, a formal agreement
between the two firms covered steel supplies from Stewarts and
Lloyds to TI, the joint purchase of two key competitors, Bromford
Tube and Howell, co-operation on research and development, a
demarcation agreement to avoid competition and, where there
were overlaps, to fix prices and, if necessary, quotas, and an
exchange of shares. While TI announced, 'We have converted a
potential menace into a real bulwark', Stewarts and Lloyds had
further secured itself against attack from inside the camp while
pursuing its backward forays into iron and steel, although a
promising area of future forward diversification had been closed.
The liaison with TI was exceptional in being at least partially
public. Others were kept secret, for example an agreement with
one of the more significant 'independents', the Wellington Tube
Works, concluded in 1934. Here Stewarts and Lloyds used its
fast-strengthening position as a cheap steel producer, able to
supply on specially favourable terms, to tie Wellington into a
complex fifteen-year agreement on prices, market share quotas
and selling policy.[14]

(3) *Leadership tactics in the domestic tube trade* largely
related to the international competitive struggle. Basically, Stewarts and Lloyds sought a tight-knit domestic cartel partly to
prevent price attacks before it became impregnable as a low-cost
producer and to strengthen safeguards against new entry; and
partly to reduce imports of cheap foreign steel, on which many of
the independents relied, and to ensure a strong bargaining posi-

tion with international tube interests. Price agreements within the domestic trade had had a chequered history, although by 1906 they were already stronger than in most of iron and steel. Between 1921 and 1932 frequent changes in discounts of up to 10 per cent upwards or downwards, depending on the trade situation, were agreed for most products. But with more than a score of separate manufacturers, mobile entry conditions and a tail of mavericks, discipline could easily be disrupted.

To cope with this Stewarts and Lloyds used a variety of tactics: efforts for exclusive supply agreements between the relevant trade association and the railway companies in the case of loco tubes, designed to cut out a particular maverick; threats to resign from the British Tube Association unless 'a small minority of outsiders' joined a price agreement, combined with private plans to take one or two of them over; threats to reduce prices unilaterally, in the case of one group of products by 7½ per cent, unless a minority of dissidents came into line. Slowly, the firm's grip tightened, helped by Corby, so that by the mid-1930s its power over the remaining small independents had become increasingly institutionalised. Indeed, the long-perceived political advantages to it of deliberately maintaining this sector increased (see chapter 9).[15]

(4) *International operations.* For many years Stewarts and Lloyds had sought, alongside others, to strengthen international cartel arrangements. These involved both the Continental tube manufacturers, among them the formidable German firms, Rheinrohr and Mannesmann, and the Americans. As competition intensified in the late 1920s, the agreements were threatened even more by inter-firm national rivalries than by domestic dissidents. One painful result was an inflow of cheap imports, particularly of the standard, 'bread-and-butter' types of tubes. At the same time the rising tide of economic nationalism was putting up tariff barriers even in the important Dominion markets. Stewarts and Lloyds responded to these threats in three main ways. First, it secured its long-established position in the old Dominions. In particular, selling operations in South Africa and Australia, the former generally profitable since the 1900s, the latter much less so, were urgently recast and extended to embrace large-scale local manufacture of tubes. Second, from 1929 onwards Stewarts and Lloyds threw itself into a struggle to maintain the international cartel. At first tighter control of the domestic trade and a vigorous use of both diplomacy and the clenched fist towards overseas interests only partially made up for the continued lack of tariff protection and only partially diminished the flow of imports. It required the imposition of tariffs in 1931–2 to rectify

this position. Third, when the cartel was threatened again in 1935, this time by intensified German nationalism, Stewarts and Lloyds had worked up sufficient muscle, with tariffs *and* Corby behind them, to retaliate with vigour. After two years of 'insensate competition' the Germans 'sought an armistice' (Macdiarmid's phrases).[16]

All of these achievements, however, hinged on the lynchpin strategy of creating a works producing iron, steel and tubes which would be large, up-to-date, fully integrated and optimally located. It was this which was to be the acid test of managerial drive and skill. For whereas the firm was used to domestic dominance and the international top table within the limited world of steel tubes, the plan to move from the lower échelons of the iron and steel industry to its top league required a sharper adaptation. It meant invasions, captures and displacements of a more difficult kind, national prominence, a more exalted *haute politique*, confrontations with the Bank of England, Whitehall, the NFISM and the BISF, and with the existing great powers in iron and steel, not least the USC.

(5)   *Iron ore acquisitions*. Stewarts and Lloyds had come to possess a foothold in Northamptonshire by acquiring Lloyds Ironstone Company (LIC) in 1920 (see p. 53). As it increasingly realised this area's potential as a source of cheap, high phosphorus iron ore, suitable for making basic steel in large quantities and highly economically, the firm embarked on an extensive programme of corralling the relevant resources in the area. Several factors helped: local knowledge as a result of the Lloyd family's longstanding interests in the area even before 1920; greater financial ability to acquire iron-making concerns and land interests other than iron and steel firms; freedom from heavy commitments in other areas.

In autumn 1928 S. J. Lloyd reported 'all the information he can collect regarding Northamptonshire Iron Reserves, Blast Furnace Plants, etc.' (later adding several substantial surveys). Thenceforward the invasion accelerated particularly in so far as technical assurances of ore suitabilities gathered weight. In 1928 Stewarts and Lloyds estimated that it controlled 106 million tons of Northamptonshire ores; by 1930 this was 450 million tons (and 26,000 acres); by 1932 it was 500 million tons. The handful of existing local iron companies, apprised of Stewarts and Lloyds' desire to eliminate competition and its 'fear of an outside firm absorbing any of the Northamptonshire undertakings and substantial Ironstone Reserves', tried to exploit their position but not very successfully. In 1930, £315,000 was paid for an Oxfordshire-based firm, the substantial Islip Iron Company, and a further £152,000 for two smaller firms, Bloxham and Whiston, and Newbold. These prices were probably relatively moderate, doubtless reflecting the firms' inability to exploit their iron ores fully and hence to pose a serious competitive threat by going it alone. Stewarts and Lloyds' successful invasion was stealthy as well as quick. As Macdiarmid confessed to Montagu Norman in January 1930, the firm's consequent potential for producing steel 'considerably cheaper than anywhere else in the U.K.' was 'not generally known, and it is a fact which we do not wish to be too well known.'[17]

(6) *Plans for Corby.* Meantime the Stewarts and Lloyds board had several times considered Corby, the existing LIC centre, as a site for the new manufacturing complex, and in 1929 this idea gathered strong technical support. Macdiarmid met H. A. Brassert, the well-known German-American consultant who was currently advising the leading Scottish steelmasters. He was impressed by Brassert's knowledge and flair, just as Brassert, equally impressed by Macdiarmid, became fascinated by the Northamptonshire project. Under the terms of a deal completed by October 1929, Brassert was to report on Stewarts and Lloyds' economic and technical position and also to act as a consultant for three years without helping any other interests with existing or proposed investments in tubes. Brassert's proposals were to be a major turning-point. By spring 1930 he was arguing cogently for Corby as the fulcrum and advocating a project of historic proportions in the iron and steel industry. The iron, steel and tube plant suggested would cost upwards of £6 million with an annual steel capacity of over 600,000 tons, mostly for tubes but with a large proportion to be sold outside.

As the implications of this gigantic proposal sank in the Stewarts and Lloyds directors showed distinct qualms. They had technical doubts, for example over transport costs, fuel supplies and ore suitabilities for certain steel-making processes. More important were three strategic questions. (1) Was it wise to press ahead with such a project while there was still uncertainty about tariffs? At least, as one director put it, 'the powers that be should know that we go in on the expectation that long before the plant is in commission the safeguarding of steel in this country will have become an actual fact'. (2) What about the risks of trouble from other large steel producers *vis-à-vis* the non-tube production proposed? And (3), Was it wise, anyway, to assume that Stewarts and Lloyds could never again buy in outside steel supplies at reasonable prices, as it had done before 1914? Of these questions (1) missed the economic cogency of the plan even without tariffs, (2) foreshadowed a genuine problem, whilst (3) cast doubt on the entire strategy by implying, unjustifiably, either that cheap foreign supplies should be relied upon or that the rest of the iron and steel industry would be able to reorganise quickly to produce economic supplies. In the longer-term, though, relative to other expansive things that the firm could have done, the issues raised by (3) were clearly fundamental (see below). In the event the strategic worries soon vanished as Macdiarmid and the board grew increasingly enthusiastic. Before long their main anxiety was how on earth to raise the money.[18]

(7)   *Fund-raising and public sponsorship.* Starting in early 1930, Macdiarmid embarked on an effort to obtain finance from the BID Co.[19] Some two and a half years were to elapse before he succeeded. Major problems arose because the financing of a single company was exceptional for the BID Co, because of economic and financial constraints, particularly through 1931, and because of the ambitious character of Stewarts and Lloyds' demands. Basically, what Macdiarmid sought was finance for a project which would complete his firm's mastery in tubes, consolidate its hold on the ultra-cheap Northamptonshire iron supplies for steel, and enable it to become a significant general steel-maker. Moreover, he wanted the money on specially favourable terms, on a fixed interest basis that would leave Stewarts and Lloyds all the equity, and preferably also on conditions that would prevent other steel firms from horning in on the tube trade. Of course, the project was highly attractive to the public interest thinking of the time, as soon became clear to the BID Co. It offered large degrees of import substitution pending a possible tariff, export potential, and bargaining power *vis-à-vis* the foreigner in the international tube cartel, along with the rationalisation of an important trade and the exploitation of a national asset. But it was still true that Stewarts and Lloyds sought an exceptional degree of favoured treatment and, in effect, of public support for a substantial advance of its power, which many others, including important interests in iron and steel, would resent.

By 1931 the firm was resorting to a rich variety of tactics in its efforts to influence the BID Co. It resumed previous flirtations with American financing sources and went to the government behind the BID Co's back, complaining that 'the only difficulty' was the latter's 'quite impossible terms', thus infuriating Charles Bruce Gardner. It got trade union leaders to plead on Corby's behalf. At one point it even played on some Labour ministers' flirtation with steel industry nationalisation. Macdiarmid claimed that Graham, President of the Board of Trade, had asked 'why the whole business should not be treated as a public utility', something 'that the Government could probably arrange'. Disingenuously, Macdiarmid told the BID Co that 'so far as his Company is concerned, this arrangement would suit very well', even though he took care to add a ritual condemnation of 'socialisation...from the broader point of view'. At various times Stewarts and Lloyds argued for £½ million reimbursement for their spending on Northamptonshire iron acquisitions; 'very, very special interest rates' (Bruce Gardner's phrase); financing conditions which led one BID Co director, Sir Guy Granet, to exclaim, 'I do not see why Stewarts and Lloyds should get all the jam'; limitations on the amount of Corby steel the firm would take for tube purposes; and a stipulation, surprising in view of its commitment to the scheme, that it would only take steel of superior quality resulting from it. On its side the BID Co increasingly resorted to delaying tactics. If these partly reflected the financial constraints of 1931–2, they were also designed to cut down Stewarts and Lloyds. Macdiarmid showed considerable nerve during this period when, as he put it later, raising the funds 'remained a distant and sometimes a retreating hope'. Only in September 1932 did a package finally emerge. This provided £3.3 million of outside finance (Union Bank of Scotland, £1.5 million; Lloyds Bank, £1 million; Prudential Assurance, £250,000; BID Co itself, £550,000).

How far was this a Stewarts and Lloyds' victory? On one key issue the firm had had to climb down. In Macdiarmid's later words, from 'a full-blooded' project 'the scheme was narrowed down to a purely Stewarts and Lloyds affair, based on the bare necessities of survival against free competition from the Continent'. It was to be smaller, and basically restricted to tube industry outlets. The cutting-down derived from financial stringency, and also from counter-lobbying by steel industry interests which feared a low-cost, super-efficient invasion, based on Corby, into their own markets. Hence the full economies of steel production would be under-achieved. The terms of the finance package were a mixture: low interest rates but a short repayment period.

As against all this, however, Stewarts and Lloyds had registered considerable gains. Its repeated insistence on 100 per cent equity ownership had not even been questioned. Virtual monopoly of the British steel tube industry had been largely ensured. The firm was enabled to jump ahead of others with a large investment project when construction costs were low and, as Macdiarmid put it later, 'We were able to create keen competition for every order'. Although he would have preferred 'complete freedom in development', a demarcation agreement with some important steel interests stipulated that if Stewarts and Lloyds kept out of iron and steel markets, they would keep out of tubes (see below). Arguably, though, with its consolidated grip on cheap iron ore sources and super-cheap steel-making, Stewarts and Lloyds' power to threaten them was potentially the greater. Not least were the advantages of quasi-official sponsorship. The firm's position as a showpiece rationaliser and national economic champion was recognised by the BID Co support, as also by Whitehall, the City, the press and trade union leaders. All this occurred under a new fiscal régime whereby tariffs, introduced in spring 1932, were closing off import competition and providing greater security to both the tube trade and the iron and steel industry. Moreover, there were virtually no attendant strings *vis-à-vis* public investigation of Stewarts and Lloyds' subsequent behaviour, its pricing practices or its contribution to collective responsibilities in iron and steel (see Chapter 9).

With Corby as the pivot, Stewarts and Lloyds proceeded to expand hugely. The new plant's construction proceeded in 1933 and 1934. Its birth pains were probably no worse than the average. A complementary rundown and relocation of many other activities was carried through speedily (for its managerial and social implications, see Chapters 7 and 8). The effect of all this, together with tariffs and economic recovery, was a huge boost to the firm's profits. If anything, these were understated partly because of conservative profit estimation and declaration, partly because depreciation provisions shot up (see Table 6.1).[20]

This brief account has obscured an important aspect of the growth process – the extent to which growth strategies were conceived as a whole. Throughout the process Allan Macdiarmid's overall strategic thinking is evident. In his papers to the board the background informational inputs to decision-making (detailed and quantified as before, in the firm's tradition) are firmly subordinated to a sweeping, carefully deployed, often elegantly expressed approach to big policy issues. One component of this is a keenly modulated sense of the complexity of the external environment. For example, 'The position of Stewarts and Lloyds, not directly on account of Protection, because they had protected their position earlier by international agreement, but indirectly so, in respect that Protection resulted in a big revival of the home market, has also completely changed' ... 'It would require an incredible revival in world trade to enable us to keep the new works and all the old Works going and against the current of such a revival the withdrawals due to economic nationalism – Australian tube works etc., – are always setting in.' Macdiarmid's distaste for limitations on Stewarts and Lloyds' freedom of action is palpable. For instance, without demarcation 'there is little chance of obtaining finance either through the BID or the Banks' but even then 'it would appear inadvisable...to bind Corby absolutely not to supply anything but tube steels' (1932) and 'We must hold Corby and Hickmans absolutely free from outside control or restriction' (this by 1935).

Forward strategic movement is to be rapid but not precipitate. It should be possible 'to carry through the transition in a series of steps which, while completed as rapidly as possible, will also be devised to provide time to readjust the general organisation [of the industry] to the changing conditions' (1932). Moreover, that forward movement is repeatedly viewed in military fashion as with a series of campaigns. The battle is conceived partly in defensive terms: 'the danger of attack on our tube interests by steel manufacturers', 'our proper defence is attack', 'threats of encroachment on our own trade', 'the necessity to keep our hands free to retaliate against the aggressor in his own particular line', 'a fight outside the tube terrain'. But the moves actually contemplated or taken are often anticipatory or pre-emptive. Thus major expansion moves are planned in late 1935 because 'if we stay as we are, the advantages in our favour will gradually diminish. The expenditure which the other makers are freely incurring will bring their costs nearer to ours. The matter brooks of no delay.'[21] Over and over again the implication is that business is a war game in which Stewarts and Lloyds must always keep several moves ahead (for the collective-industrial and public policy implications see Chapter 9).

Stewarts and Lloyds' notable success story of growth involved some internal conflicts as well as pain and error. As we have seen, strong arguments arose initially over Corby in 1930. During the awkward interval between the board's final commitment to Corby in early 1931 and the clear assurance that finance would be forthcoming more than a year later, anxieties clearly existed, at one point described, at least for outside consumption, as 'desperate'. In so far as the firm was faced with a variety of acquisition possibilities, difficult choices arose. In 1928–32 in particular, when funds were relatively straitened, a constant juggling act had to be performed between two acquisition sectors, the Midlands iron ore fields and the tube trade, each providing multiple choices. Signs of explicit acquisition priority criteria (size, ten-year profit capitalisation, nuisance potential, etc.) do not eliminate the likelihood that these choices involved much heart-searching. Moreover, a number of outright mistakes lurk near or just below the surface: high-pressure tactics towards the BID Co that rebounded, serious underestimates of the Corby infrastructure costs, a probable initial tendency to underestimate the strength of steel industry fears of the project.

Speculation about alternative growth possibilities and missed opportunities is more difficult. For example, should Stewarts and Lloyds have diversified into smaller and finer tubes, substantially the area occupied by TI? Should it have followed a different course by later competing with and/or taking over TI? Again, should the firm have moved into tube products made of other types of metal (zinc, copper, etc.)? Whether such downstream and substitute-product diversification would have served the firm better than its massive invasion of iron and steel is a fascinating issue. In the longer term, perhaps, faster growth, more scope for the firm's commercial brilliance and fewer political risks might well have resulted. To a rational man in the late 1920s and early 1930s, however, such paths would have seemed tenuous, even supposing they were glimpsed. Both the firm's traditions over several decades and the current exigencies suggested that long-standing objectives should be brought to a climax in the shape of Corby. To modernise and expand one's existing products and existing supply sources appeared to be most reasonable. In this sense, for all its brilliant execution, the strategy was basically conservative, although in retrospect it is hard to see how Stewarts and Lloyds could have obtained comparable growth in any other way, at least up to 1939.

## Growth muffled: the USC

The USC's growth record in the 1930s stands in striking contrast to both Dorman Long's and Stewarts and Lloyds'. In a sense it lies between the two extremes they represent. For the USC's growth performance was essentially moderate and respectable, but unexciting. Existing resources were more fully used and there was some diversification within existing markets. The firm grew at a reasonable rate relative to others, thanks partly to the economic situation, and partly to a strengthening of management skills. On the other hand, a number of significant growth opportunities were mishandled or ignored, and in some more abrasive fields of competition there was outright failure.

The volatility of profits makes them a poor measure of growth. The USC's profits recovered dramatically through the 1930's but from a far lower base-point, at the trough of the recession, than Stewarts and Lloyds' (see Table 6.1). A long-term comparison between the USC's 1937–9 average profits and those achieved by its constituent companies in 1914–18 (as estimated in 1920) suggests little profit growth in real terms. The USC's production of steel ingots increased rather less than the total for Great Britain between 1931 and 1937: an admittedly crude measure which ignores some shift in the firm's production 'mix' from heavy to lighter, more specialised and higher value products. The firm estimated in 1937 that its tonnage share of UK steel production increased by one or two percentage points between 1929–30 and 1935–6, only a slight improvement. External sales nearly doubled from about £7.5 million in 1931 to about £14.9 million in 1939, probably not much above the industry average. Total capital spending on new plant and equipment and acquisitions over the whole period was £10.8 million. As a crude measure this was somewhat less than the firm's 1930 capitalisation, whereas in Stewarts and Lloyds the combination of Corby, the tube takeovers, iron acquisitions and the rest, represented spending well above one and a half times its 1930 capitalisation: a grand total of £15¾ million.[22]

*Table 6.1   Profits in the 1930s (£,000s)*

|       | 1930 | 1931 | 1932 | 1933 | 1934 | 1935 | 1936 | 1937 | 1938 | 1939 |
|-------|------|------|------|------|------|------|------|------|------|------|
| USC   | n/a  | 191  | 345  | 503  | 1095 | 1239 | 1251 | 1594 | 1977 | 1426 |
| S & L | 591  | 317  | 272  | 616  | 1005 | 801  | 1192 | 2303 | 2440 | 1753 |

*Sources*: USC: Andrew and Brunner, op. cit., p. 183, trading profits after taxation and interest; S and L: S/CB, net profit after taxation and interest, *viz.* in both cases before provisions for depreciation, reserves and dividends.

Growth mainly meant improvements in plant and equipment within existing activities, and increased capacity utilisation. A 1929 report by Brasserts set the keynote with an emphasis on re-equipment in all the main branches. In 1930 and 1931 investment continued to be held back and great efforts were put into finding outlets for existing products: a case of unfinished materials 'crying out for the discovery', as Walter Benton Jones put it, 'of finished or semi-finished products to absorb them'. Efforts to bring about rationalising mergers in West Cumberland in 1930–1 came to grief despite support from the BID Co. Such limited acquisitions as occurred were mainly designed to round out existing interests as when the NLIC was bought from Stewarts and Lloyds in 1931 because of its closeness to the Frodingham works. Again, the USC's technically sophisticated diversifications into the stainless steel trade and electric steel-making were a logical development of its special steel activities in the Sheffield area.

When substantial expansion started in 1934 this represented a combination of (1) rapid moves down the short-run average cost curve as economic recovery increased output within existing capacities, and (2) the coming-into-effect of productivity improvements from the new régime's efficiency policies (see Chapter 7). During the investment boom of the late 1930s the USC's capital expenditure increased rapidly to around £1 million per annum in 1936–7, then to £2.4 million in 1938, and £3.1 million in 1939, forming an increased share of the industry's total. This basically reflected the company's size and efficiency orientation rather than any aggressive bid for strategic expansion. Much of the investment was of a 'catching-up' type; it concentrated on the firm's existing locations and operations, and it carried no overtones of a major offensive.[23]

It was in the 'Midland corridor', according to Allan Macdiarmid 'the most important steel region in the country', that the USC met the biggest test with regard to strategic growth. In this context the 'Midlands' was an ill-defined, inverted and lopsided triangle stretching from Lancashire to Lincolnshire and including both Oxfordshire and the important Northamptonshire complex to the south. By the early 1930s five principal groups operated in this area, all of them major users of Midlands iron ores. *Richard Thomas*, the large tin-plate firm, possessed the important Redbourne works in Lincolnshire, although by 1935 this company's plans for further concentration in South Wales and the individualism of its chairman, William Firth, were causing it to recede. The *Lancashire Steel Corporation* (LSC) was a new, medium-sized amalgamation, sponsored by the BID Co as a rationalising move and still partially financed by them. The *Stanton Ironworks*, a private company, specialised in cast iron pipes and possessed substantial iron ore and iron-making facilities, including two works in Northamptonshire. Finally, of course, there were *Stewarts and Lloyds* with its established Midland interests, its new strength in Northamptonshire iron ores and its fast-growing complex at Corby, and the *USC* whose interests in Lincolnshire (Appleby and Frodingham), in South Yorkshire coal and coke, and in the Midland markets for semi-finished and finished steel products, were so important to it.

Not only did substantial direct competition already exist between these firms. New frictions soon emerged with regard to the mass production possibilities opening up particularly in Northants and with regard to the potential these offered, in turn, for cheaper semi-finished products, especially for the fastest-growing markets which were also, to a substantial degree, in the Midlands. A tendency followed on the part of relatively detached parties, notably the BID Co and later the BISF under Sir Andrew Duncan, to support moves for pacification, co-operation or even fusion. The consequence was a series of threats, bargains and attempted integrations, of fast-shifting alliances and feuds, as the 'Midland corridor' became a scene of industrial power politics.

The USC's worries about Stewarts and Lloyds were already evident in 1930. Corby seemed likely to produce steel more cheaply than the USC's own cheapest works, Frodingham. If Corby were to be of an economic size for iron and steel, its outputs would greatly exceed Stewarts and Lloyds' requirements for tube steel. Hence the latter firm could supply wider outlets for steel tubes which were already served by the USC. Worse, it could invade general steel-making at a time of continued recession, cutting into the USC's markets, particularly in the Midlands, and cutting across the firm's own rationalisation plans. All this with semi-official support and Bank of England money. These worries eventually carried sufficient weight with the BID Co, anxious to be impartial and increasingly financially constrained, for it to engineer a *modus vivendi* between the USC and Stewarts and Lloyds, finally achieved in 1932. As already explained, the Corby project was cut back. In addition, the two firms formed an agreement. Publicly, this stipulated co-operation on sales policy, technical matters and 'unnecessary duplication of plant'. Privately, it included agreements on the basis of 'a long period: viz. 30 years', that the USC would not compete with Stewarts and Lloyds in the tube trade, that it would supply Corby with certain materials, that it would market any Corby steel in excess of tube trade requirements, and, very important, that if Corby's subsequent enlargement and the further development of Northamptonshire as a general steel centre were 'called for in the national interests', the USC would 'have the right to a 50/50 partnership with Stewarts and Lloyds on terms to be mutually agreed'.[24]

This agreement seems to have worked fairly well for at least three years. Liaison committees arranged for exchanges of supplies and technical information; expansion plans appear to have been co-ordinated. By 1935, though, strains were emerging. For one thing, Stewarts and Lloyds were privately meditating a more generally bellicose stance. In a particularly important board paper in January 1935, bulging with military metaphors, Allan Macdiarmid bitterly regretted that Corby was not being 'exploited in the full National interest' while at the same time South Durham and other large firms were 'threatening invasion of our particular province' [the tube trade.] His key point was that Stewarts and Lloyds should quickly move into a state of readiness for major expansion at Corby, avoiding both trade association and Federation restrictions (now imminent) and pre-empting its rivals and the potential invaders. Various alternatives were being studied – 'ranging from the most ambitious', a large-scale expansion to reduce imports and make more profit for Stewarts and Lloyds 'on a considerable outside sale of steel', to 'the most modest', a minimum plan to defend the firm's tube trade position by 'enabling us to attack would-be invaders in their own territory'.[25]

An increasing chasm was soon pointed up by the USC's own moves. By autumn 1935 it was engaged in merger negotiations with the LSC (see below). Walter Benton Jones simultaneously pursued an idea with crucial overtones both nationally and for Stewarts and Lloyds: basically, a proposal that all five of the key firms mentioned above should form a consortium, based on 'equal partnership'. This would keep out newcomers, develop the Midland iron ore fields, and co-ordinate commercial and investment plans. Benton Jones probably found the idea appealing because it would safeguard the USC's interests in Northamptonshire, contain Stewarts and Lloyds more effectively within a wider grouping, and accord with public interest thinking. Later, Sir Andrew Duncan and the BID Co were to find equally attractive the idea that Stewarts and Lloyds' ore deposits and those of Northamptonshire should be developed in the national interest by 'a joint enterprise'. But although Benton Jones sedulously pushed the idea for many months, his hoped-for partners appeared less keen. An investigation by Brassert, to which they all agreed, postponed the crunch well into 1936. For Macdiarmid, who was currently pursuing detailed separatist plans, the proposal was no doubt especially distasteful, although perhaps a useful means of playing for time. In his view, Stewarts and Lloyds' advantages sprang from many years of superior enterprise, foresight and effort. Why should its hard-won gains be horned in on by others merely jealous and less meritorious? Why should his firm be shackled just when new opportunities were opening out? By now the stage was set for all-out war between the two giants.

In April 1936 Macdiarmid suddenly told Benton Jones that 'owing to substantial changes in the situation during the last four years', Stewarts and Lloyds was withdrawing from the alliance, and also from the more recent consortium proposal. The USC felt it had been cheated not only by Macdiarmid but also by Brassert whom it criticised both for withholding key data and for producing 'evidently prejudiced' findings in favour of Stewarts and Lloyds. In the bitter conflict that followed, both sides quickly lodged complaints with the BID Co. The USC said their ex-partners had been high-handed and 'cavalier'. More stridently, Macdiarmid described the USC as 'greedy, unaccommodating, and almost stupid', and alleged that it wanted to control the proposed new grouping and even Corby itself, a charge which Robert Hilton hotly denied and which was not borne out by Benton Jones's 'equal partnership' scheme. More important were the two firms' decisions. If Macdiarmid's plans were already far advanced, the USC was ill-prepared. In a sudden rash of emergency moves it decided to raise large-scale new capital (for a 'War-chest', as Bruce Gardner aptly put it). Major expansions in Lincolnshire or Northamptonshire were hurriedly investigated. Senior officials were despatched overseas to enquire into the latest developments in sectors which it was feared Stewarts and Lloyds would invade.[26]

In the final scenes of the drama which now followed the USC tried to get its own back on Stewarts and Lloyds and to wage an offensive, but largely failed. One by one it discovered that the avenues conceived as promising were closed off or would shortly become so, usually by its rival.

(1)  An approach to the Stanton Ironworks to build a new, joint iron and steel works in Northants was the first move: 'they bring us ore, they may be able to bring us pig iron...they may ultimately bring a market for steel and because they are not yet steel manufacturers they will accept leadership from us.' Benton Jones wooed Stanton with the idea that it 'must in self-defence [against Stewarts and Lloyds] make steel pipes sooner or later'. His bid failed mainly, no doubt, because Stanton's controllers were playing for bigger stakes. Later, in 1939, Stanton was taken over by Stewarts and Lloyds, a move still further consolidating the latter's Northants position.[27]

(2)  Discussions about a merger with the LSC were urgently renewed. Repeated efforts on this front since 1931, strongly encouraged by the BID Co, which was anxious to extricate itself from the LSC, had already failed. A USC offer of about £4 million in autumn 1935 was not enough for the BID Co to get its money back and the LSC's strong-willed managing director, John E. James, disliked the USC. So it was hardly surprising that yet another USC bid should fail. More frustrating was the spectacle of the LSC, 'attracted by Stewarts and Lloyds' works and by Macdiarmid personally', increasingly flirting with them. Even the BID Co, anxious to prevent the LSC going over completely to Stewarts and Lloyds, could not stop the two firms reaching an important agreement in October 1937 for the LSC to build plant at Corby, to be supplied by Stewarts and Lloyds. This pact rescued the latter from the invidious position of expanding Corby com-

pletely on its own. A further major expansion of Corby followed, feeding a particularly up-to-date LSC development in precision-type rolled products for the electric and motor industries.[28]

(3)  Attempts to revive the consortium idea on still grander lines got nowhere. Montagu Norman, persuaded into peace-making efforts by Sir Andrew Duncan, was sympathetic. But Benton Jones's new proposals, this time for a full amalgamation of all the principal firms, was clearly a non-starter.[29]

(4)  The idea that the USC should build its own steelworks in Northamptonshire was investigated but this possibility had been pre-empted, largely by Stewarts and Lloyds. 'There was not sufficient unsecured ore on any one site in Northamptonshire on which a works could be installed where the carriage of ore would be eliminated.'[30]

(5)  The USC board also decided that another unilateral possibility, a large-scale entry into the wide-strip field, would be impracticable. Its risk aversion here seems to have reflected four main fears: that an economically sized wide-strip plant would be too large for the firm to supply from its own steel; that it would produce more than the market could absorb; that it would represent too big a diversification; and that it would offend others in the industry. However, the wide-strip trade did eventually provide the USC with a consolation prize for its other strategic failures. In 1938, helped by the BID Co, it went into partnership with a smaller firm, John Summers, an existing specialist in the sheet trade; an arrangement which was to give the USC a valuable entrée into this fast-expanding sector.[31]

Across the whole field of competitive growth strategy, therefore, there can be little doubt that the USC had failed. In so far as it wanted to succeed in this sense, it was no match for firms like Stewarts and Lloyds by whom it had been strikingly outmanoeuvred. The explanation of this contrast once again raises complex issues of interpretation. Clearly, an important contributory factor was the USC's bungling compared with Stewarts and Lloyds' agility. Competitive gamesmanship was lacking. There were no fall-back plans. Too much hope was pinned on the Stewarts and Lloyds alliance whose fragility was misjudged. The pursuit of alternatives which had been rushed into was too ardent and inconsistent. No doubt, too, there was something in Bruce Gardner's and Sir Andrew Duncan's view that the USC were 'bad negotiators', which was largely a reflection on both Benton Jones and Hilton.

However, all this implies a one-dimensional view of management, narrowly confined to questions of competitive skill and technique. It ignores part of the central argument of this study to the effect that business policies are fraught with conflicts of objectives and values. Within the perspective of earlier chapters, management is repeatedly involved in triumph-cum-tragedy dramas, in a bitter-sweet process of trade-offs between growth, efficiency and social action. In the main these trade-offs reflect, it is hypothesised, the stage of evolution of the firm and, more critically, its managerial preferences and abilities, and their frequent biases. According to this view, certain underlying managerial and corporate tendencies mean that weaknesses in one field may be the reciprocal of strengths in another.

Thus the USC might be defective on aggressive growth not simply because of ineptitude but also because its management were reacting dialectically against the contrary excesses of a previous phase and, perhaps more important, because they preferred other values and excelled at other things. Dorman Long's merger failure may be explained by its performance and priorities in other fields. As for Stewarts and Lloyds' victory in the growth battle, this might have been purchased at a substantial price. The question of whether such an expansionary triumph was paid for by shortfalls in efficiency or social action cannot be burked. Armed with the greater detail available on the 1930's, it is to these over-arching issues of long-run trade-offs that we return in the next three chapters.

## Notes

1  W. G. Willis, *South Durham Steel and Iron Co Ltd.* (1969). SD/DM (208/1/1/1). D/CM, particularly Report of Joint Committee, Dorman Long, South Durham and Cargo Fleet, March 1930, for details of 1929 financial structures, and later report from Peat, Marwick, Mitchell for 1929 outputs.
2  D/CM, Report of Joint Committee, March 1930, op. cit. D/DM, 8.7.30. BE, SMT 3/107, Notes for Mr Campbell, 5.6.30; SMT 3/18, C. Bruce Gardner, discussion with A. Dorman, 24.7.30. SMT 9/1, 21.3.32, 10.4.33. PRO/BT 56/37, Sir H. Mensforth to Sir H. Wilson, 25.11.30. D/CM, Report from Peat, Marwick, Mitchell, August 1932.
3  PRO/ BT/ 56/21: memorandum to Sir H. Wilson, 14.3.30; Sir H. Wilson, discussion with Sir W. Larke, 8.7.30. DT 56/37, Memorandum from J H. Thomas to the prime minister, June 1930. CAB/CP 420(30), Memorandum by President of Board of Trade, 12.12.30. *The Times*, 28.11.30. BE: Report on the structure of the iron and steel industry of Great Britain by C. Bruce Gardner, December 1930. For the classic statement that reorganisation must precede tariffs see PRO/ CAB/CP 189(30), Economic Advisory Council, Iron and Steel Committee (Sankey) Report.
4  D/DM. D/FC. SD/DM. BE: SMT/2/93, SMT 3/18, SMT 6/4, SMT 9/1, BID 1/ 79-87. PRO BT 56. D/CM, E. Furness to B. Talbot, 20.1.31.

5   D/CM, memorandum to Viscount Furness, 14.4.30. Meeting, E. Furness and C.
    Mitchell, 19.1.31. D/FC, 12.6.30. BE: SMT 2/93, E. A. Peacock to Montagu
    Norman, 25.2.31; BID 1/81, C. Bruce Gardner to E. A. Peacock, 10.3.31 and
    9.4.31: BID 1/82, Meeting, 23.12.31.
6   D/AGMs, 12.12.31 and 21.12.32. D/DM. D/CM. BE: SMT 9/1, 10.4.33. SMT 3/
    18, Interview, C. Bruce Gardner and B. Talbot, 23.5.30. BID 1/82, Memoran-
    dum, 19.10.31; Sir W. McLintock to C. Bruce Gardner, 21.10.31; H.A.B.(?) to C.
    Bruce Gardner, 5.11.31; A. N. McQuistan to A. McColl, 1.4.32.
7   BE: SMT 2/93, especially E.A. Peacock to M. Norman, 25.2.31; SMT 3/18,
    especially C. Bruce Gardner, interview with B. Talbot, 23.5.30; SMT 6/4,
    especially meetings with E. Furness, 15.7.30 and 28.7.30, Sir W. McLintock to
    N. L. Campbell, 5.8.30; SMT 9/1, 11.5.31. PRO BT/56/37, C. Bruce Gardner to
    Sir H. Wilson, 2.12.30. D/CM. SD/DM, 30.9.31, 3.11.31, 19.5.32. DL and SD,
    Letter to Shareholders, 30.6.33 (234/1/33).
8   BE: SMT 9/1, 23.6.30, 18.1.32. SMT 3/18. BID 1/81, 82, 83.
9   D/DM, 18.2.30, 15.11.32. D/FC, 13.3.30, 11.6.31, 16.3.33. SD/DM, especially
    30.6.32. D/CM, B. Talbot to E. Furness, 27.8.32, B. Talbot *et al.* to E. Hunter,
    30.5.33, G. Sturt to B. Talbot, 4.12.22. BE: BID 1/82, F.C. Goodenough of
    Barclays to C. Mitchell, 5.11.31, A. N. McQuistan to C. Mitchell, 10.11.31, F. C.
    Goodenough to C. Mitchell, 31.3.32, F. C. Goodenough to C. Bruce Gardner,
    19.5.32.
10  BE papers as above. D/DM, D/CM, including protest letters from shareholders.
    DL and SD letters to shareholders, 30.6.33. SD, meeting of debenture holders,
    and meetings of ordinary shareholders, 19.7.33. DL, meetings of ordinary
    shareholders, preference shareholders and debenture holders, 27.7.33.
11  Chairman's conference, S and L, May, 1962, historical survey by A. G. Stewart.
    S/BM, 9.3.28, 11.2.29, 24.4.29, 7.6.29, 25.10.29, 29.11.29, 1.4.30, 6.5.30, 3.6.30,
    1.10.30, 11.11.30, 25.9.34. Miscellaneous material and cartels (1791/1/21).
    Proposals for reorganisation of British Tube Industry, September 1932. F.
    Scopes, *The Development of Corby Works* (1968), pp. 55–6. John Spencer Ltd.
    (065/2/3). Scottish Tube Company AGMs, 1919–31 (1791/6/16).
12  S/BM, 21.9.16, 25.7.24, 7.6.29, 29.7.32, 26.11.35, 31.12.35. S/GPC, 26.3.29,
    23.4.29, 6.6.29, 2.6.30. S/MC, 10.2.32, 26.7.32, 6.9.32, 27.9.32, 10.1.33, 21.2.33.
    1791/1/21, Proposals for reorganisation, etc., *op. cit.* S/AGM, 13.5.36. BE: SMT
    2/148, correspondence on S and L and British Mannesmann, 1935.
13  1791/1/21, Proposals for reorganisation, etc., *op.cit.*
14  S/BM, 1.4.30, 3.6.30, 1.10.30. S/GPC, 26.1.31. S/AGM, 6.11.30. Scopes, op. cit.,
    pp. 56–7. Tube Investments, AGMs, 6.12.28, 12.12.29, 4.12.30, 3.12.31, in *The
    Times*. 1791/1/21, Agreements between S and L and Wellington Tube Works.
15  1791/1/22, Tube Makers' Association, Papers, 1879–1931, including early
    price agreements from 1879 and trade press cuttings for 1900's, J. C. Carr and
    W. Taplin, *History of the British Steel Industry* (1962), pp. 261–2. S/OC,
    23.6.21, 8.9.21, 12.1.23, 9.3.23, 7.3.24, 2.5.24, 29.1.25, 17.7.25, 21.8.25,
    21.10.26, 14.6.28, 24.5.29, 7.6.29, 24.7.29, 6.5.30, 21.3.31, 2.2.32. For Stewarts
    and Lloyds' tactics see particularly S/GPC, 2.6.30, 1.9.30, 31.3.30, 5.5.30,
    2.6.30, 1.9.30, 15.12.30, 27.4.31. S/OC, 1.7.30, 29.7.30, 11.11.30; and Meeting
    of Large Tube Makers, 23.1.29.
16  S/BM, 8.9.21, 8.5.25, 5.6.25, 4.5.28, 6.5.30. S/GPC, 5.2.19, 12.3.19, 5.5.30,
    28.7.30, 1.9.30. S/OC, 11.3.26, 21.10.26, 29.3.28, 19.12.28, 5.3.29, 24.4.29,
    17.12.29, 11.3.30, 10.11.31, 15.12.31. 1791/1/23, Memorandum on overseas
    companies 1946. 1791/1/33, S and L (Australia) Ltd; Minute Books, including
    financial results, 1913–3. 1791/1/35, S and L (South Africa) Ltd. Minute Books
    from 1905, including financial results, 1916–26, and details of reconstitution
    in 1927. Carr and Taplin, *op. cit*, p. 489. Chairman's Conference, S and L,
    May, 1962, historical survey by A. G. Stewart. Scopes, op. cit., pp. 164–

7, 222–4. A. C. Macdiarmid's chairman's speeches as reported in *The Times*, 1929–39.

17  S/BM, 24.4.29, 7.6.29, 25.10.29, 1.4.30, 6.5.30, 29.7.30. Scopes, op. cit., pp. 64–6. Islip Iron Company, balance sheets, directors' reports (002/1/10). Kettering Iron and Coal Company, Minute Book (065/3/4). BE: SMT 2/148, Secret aide-mémoire for A. C. Macdiarmid's use in talk with Montagu Norman, 21.1.30.

18  S/BM, 24.7.29, 1.10.29, 11.3.30, 1.4.30, 6.5.30, 3.6.30, 29.7.30, 7.10.30, 16.12.30, 3.2.31. S/GPC, 30.6.30. Scopes, op. cit., pp. 54–5, 58–63, 67–79.

19  The following three paragraphs are based mainly on BE records. Of these, BID 1/100–106 contain much detail on the financial negotiations, with some supplementary material in SMT 3/73. High-level discussion and political aspects mostly emerge in some relevant papers of Montagu Norman (SMT 2/72 and SMT 2/148) and of Charles Bruce Gardner (SMT 3/18), also in various references at the Securities Management Trust's weekly meetings through this period (SMT 9/1). S/BM provide some data on both the negotiations and the parallel financial and technical problems. See also Scopes, op. cit. pp. 71–90.

20  S/BM. S/MC. S/AGM's. Scopes, op. cit., pp. 90–4, 219–21. BE: SMT/3/230, Corby progress reports; BID 1/100–106.

21  Various communications to board in Scopes, op. cit., pp. 189–90, 193, 201, 225, 227–9, 234.

22  USC, Miscellaneous historical papers, branch and subsidiaries' profits, 1875–1943. Amalgamation 1927 (006/10/3), special report by Chief Accountant, August 1927. Andrews and Brunner, *Capital Development in Steel* (1951), pp. 204–5. S and L Chairman's Conference, May 1962, historical survey by A. G. Stewart.

23  Report on the Properties and Operations of the USC by H. A. Brassert and Company, September 1929 (006/10/1). Andrews and Brunner, op. cit., Chapter 6. U/CI, notes, 30.5.31 and generally. U/MD. U/DM. For the West Cumberland merger forays see BE: SMT 7/1; SMT 3/238; SMT 2/147: SMT 9/1, various references in 1930 and 1931; SMT 3/18. BISF, Film 4. Carr and Taplin, op. cit., pp. 588–9.

24  BE: SMT 9/1, 27.10.30, 3.11.30, 17.11.30, 6.6.32. U/MD, R. S. Hilton to J. Henderson, 7.5.32 and 12.5.32, to W. Benton Jones 13.5.32. U/DM, 24.5.32, 22.6.32. For the full USC, S and L agreement see S/MC, 24.5.32. *The Times*, 2.6.32.

25  U/MD, R. S. Hilton to A. C. Macdiarmid, 4.7.32; to H. A. Brassert 24.8.32; to A. C. Macdiarmid, 3.3.34; to S. C. Lloyd, 21.3.24; to G. S. McLay, 13.10.34 and 4.1.35. U/DM, 22.11.32, 22.3.33. S/MC, 10.1.33, 21.2.33, 16.5.33, 14.7.33, 11.1.34. Scopes, op. cit., pp. 222–9.

26  U/10, W. Benton Jones memorandum 10.4.35 and his ensuing correspondence with E. J. Fox (Stanton Ironworks), Sir W. Firth (Richard Thomas), A. C. Macdiarmid and H. A. Brassert, up to February 1936. For the break-up of the USC, S and L agreement see S/BM, 7.4.36, 28.4.36; U/10, W. Benton Jones to A. C. Macdiarmid, 19.4.36, R. S. Hilton to W. Benton Jones, 27.4.36, W. Benton Jones to H. A. Brassert, 1.5.36, Note to board, 13.5.36, R. S. Hilton to W. Benton Jones, 4.6.36; U/DM, 22.5.36, and BE: SMT 9/1, 4.5.36, 18.5.36, and SMT 3/244. U/DM, 22.4.36, 12.5.36, 27.5.36. U/MD, R. S. Hilton to A. O. Peech, 27.4.35, to C. G. Atha, 23.6.36 and 2.7.36.

27  U/10, Note to board, 13.5.36 and ensuing correspondence, W. Benton Jones and E. J. Fox, May and June 1936. S. D. Chapman, *Stanton and Staveley: a Business History* (1981). Stewarts and Lloyds, Stanton merger, correspondence file (Glasgow 1207/18/12). S/CB. *The Times*, 9.7.39, 11.8.39, 22.9.39, 25.11.39.

28  U/DM, 28.7.31, 8.9.31, 13.10.31, 10.11.31, 8.12.31, 7.6.35, 25.6.35, 27.11.35,

22.1.36. U/10, Note to board, 13.5.36. S/BM, 30.6.36, 28.7.36, 29.9.36. BE: SMT 9/1, 8.6.31, 28.10.35, 4.11.35, 18.11.35, 2.12.35, 16.12.35, 13.1.36, 20.1.36, 3.2.36, 16.3.36, 30.3.36, 4.5.36, 18.5.36, 5.10.36, 19.10.36, 2.11.36. *The Times*, 1.10.36.

29  U/10, Midlands Iron and Steel Rationalisation: correspondence, W. Benton Jones and Montagu Norman, Nov. 1936–Jan. 1937. U/DM, 25.11.36, 17.12.36, 6.1.36. BE: SMT 9/6, 10.11.36, 21.12.36, 18.1.37, 1.3.37, 15.3.37. SMT 2/149. SMT 3/244.

30  For the USC's earlier interest in this idea see U/MD, R. S. Hilton to T. W. Elliott, 14.11.29. U/DM, 22.4.36. U/10, note to board, 13.5.36; Estimates of comparative manufacturing costs, Lincs and Northants (undated). U/MD, R. S. Hilton to C. G. Atha, 23.6.36 and 2.7.36. U/CC, 11.11.36.

31  U/CC, 11.11.36. U/DM, 16.12.38, 25.1.39, 22.2.39, 22.3.39, 24.5.39, 5.7.39, 26.7.39. Andrews and Brunner, op. cit., pp. 170, 206. Andrews and Brunner, interviews, 1950, W. Benton Jones and A. J. Peech. BE: SMT 2/6.

# 7 Efficiency and organisation building

As they surveyed the USC's position in January 1928, Robert Hilton and Walter Benton Jones faced a daunting challenge. First, the 'heterogeneous conglomerate', as Mannaberg had called it, was in a mess both organisationally and financially (see Chapter 5). This required courage and effort in order to effect a massive turn-round. By January 1932, four years later, although the iron and steel industry had been through even worse times and had not yet revived, the USC had pulled itself out of the mire. Management had been renewed, productivity was rising even at low levels of working, a general fall in selling prices had been more than offset by a decline in production costs, due to 'a considerable improvement in manufacturing methods, output and yield', and reduced overheads. As Benton Jones privately claimed: 'Our policy of concentrating our attention to bringing our Works to a high efficiency even in a period of general depression, has paid.' Moreover, the turn-round had occurred without substantial redundancies.

The second part of the challenge was more subtle and long-term. It was that of facing the question of what was the most efficient pattern of organisation of a permanent kind. What mixture of incentives was needed to obtain the best performance from the USC's human resources, what sort of balance between central dictation, financial rewards and community building? As the economic environment improved, could the momentum of efficiency-seeking be maintained or would increasing prosperity produce complacency and slack? Granted that the USC urgently needed more centralisation in 1928, how far could and should the centralising process go? Assuming that the concepts themselves could be satisfactorily defined, where did the optimum balance lie in the longer-term between 'centralisation' and 'decentralisation'?

## The USC's classic turnround

The first few frenetic months of the USC's rationalisation efforts in early 1928 set the pattern for much that followed. Part of the bank overdraft was funded by a new debenture issue: sacrifices were obtained from the firm's ordinary shareholders (to the extent of one-half of their equity) and its security holders (reductions in their debenture interest and claims on profits). The steady rise in bank borrowing was arrested, then reversed. In February a bludgeon was applied as general managers were ordered to reduce purchases by 20 per cent. The continuous provision of up-to-date financial information to the top command was treated as would be news of victories and defeats from the front in time of war. Stocks of materials and of semi-finished and finished products had accumulated alarmingly, and these had to be reduced. The problems of surplus stocks, pig iron costs and crippling losses were worst at the Workington Branch, and these received the most urgent attention. A start was made in cutting overheads, beginning at the headquarters where excesses were most obvious. All the time Hilton was building up a picture of the worst inefficiencies and of what would be needed to put them right. By April the main emphasis of this review swung from debts, purchases and stocks to marketing with a study of both the firm's outlets and its selling machinery.[1]

Underlying the USC leadership's approach to efficiency were certain fundamental assumptions. Rationalisation might well have taken the line of large-scale divestments and closures. A few ancillary sections could have been sold off and the previous plan to sell the Appleby works could have been revived. Although the gains would have been slow, some urgently needed cash would have resulted. The opportunity could have been seized to ditch the USC's worst financial millstones, for example the Samuel Fox colliery at Stocksbridge and, more important, the persistently crippling West Cumberland activities. It could well have been argued on strictly financial grounds that the abandonment of these trouble spots would save further losses of many thousands of pounds. Finally, there was probably scope for substantial shedding of labour in most parts of the combine. But there is no sign that any of these steps was even seriously entertained either in the critical early months of 1928 or later.

The reasons were obviously complex. A contributory factor was Walter Benton Jones's integralist, conservationist, non-radical approach, shown by his previous opposition to mergers and divestments (see Chapter 5). To this was now added the professional confidence of Robert Hilton, the new managing director, feeling that he had the ability to turn round even the most unpromising situations. An important rationale of conservation was, of course, that of taking a long-term view, of hanging on until prosperity eventually returned. When that happened, not only would the USC's Sheffield–Rotherham–Scunthorpe axis of coal, basic iron, mass produced steel and special steels, come into its own economically, subject to various improvements in the meantime. It was also reasonable to hope that plate making at Appleby would become profitable and even that the West Cumberland operations could at least be made to pay. This was essentially a concept of economic salvaging and profit improvement rather than maximising (see below). It was strengthened by a further factor which demands separate consideration in Chapter 8: a deep-seated reluctance to lay off workers and to worsen unemployment, particularly in dependant and depressed communities.[2]

Before considering efficiency promotion in more detail it is necessary to make a brief reference to an extraneous financial influence on the USC in 1929–30. This was its involvement in the affairs of Clarence Hatry, the well-known financier. Hatry's ambitious ideas for amalgamating steel industry firms in the interests of 'rationalisation', loosely conceived, had attracted a good deal of interest in establishment circles in London. After he made the (perhaps rather breathtaking) suggestion that one of his group's subsidiaries should buy up the USC, the latter's directors grew keen when it became clear that this would help towards an urgently needed financial reconstruction, including drastic capital write-downs and debt reductions. In April 1929 Hatry's terms, roughly equivalent to the current Stock Exchange market values, were accepted by a large majority of the share and security holders. Hatry largely financed the purchase with the help of City financial interests. The latter then became the majority backers of a new holding company, Steel Industries of Great Britain (SIGB), which took over ownership of the USC, hopefully as a first step towards further steel industry reconstructions. The Hatry interests themselves ran into major financial difficulties and committed various irregularities leading to Hatry's celebrated trial. As a consequence they were unable to bail the USC out of its bank debts and provide it with additional working capital, as previously promised. But this, paradoxically, was to prove a blessing. For the onus of carrying out a final capital reconstruction of the USC now fell on the broader and more respectable shoulders of the City-backed SIGB. By June 1930 a scheme was agreed by the court under which a reconstituted USC was to have an ordinary share capital of slightly over £5 million, a fixed bank loan of £1.3 million, with repayment obligations delayed until 1936, and some small continuing debentures. Its bank indebtedness and other fixed interest obligations were drastically reduced, saving the best part of £400,000 a year on interest payments and redemption provisions. At the same time the book values of the assets were vastly written down by £9.75 million to a conservative figure of some £11.5 million.

Thus Hatry's visionary scheme and subsequent disgrace, the willingness of City interests to back first him and then the USC, and the resourcefulness of the USC's directors in taking advantage of these events combined to improve the firm's financial position relatively quickly and painlessly. It should be added that the newly formed BID Co also contributed to this fortunate result in two ways: by squashing an incipient attempt by Vickers, the heavy engineering firm, to get hold of Hatry's holdings in the USC at a critical moment, and by helping the USC's standing in the City. It should also be added that when much later, in spring 1934, the SIGB was wound up, its USC shareholdings (still the only ones it had) were sold and the combine reverted to the more conventional status of a quoted company with widely dispersed ownership.[3]

One of the most striking features of the USC's efficiency seeking was the speed and range of the decisions made in 1928 and 1929. The head office was made independent and geographically distinct, and considerably streamlined. At least a dozen major managerial appointments were made from board level downwards with particular emphasis on the sales side. Centralised purchasing and costing systems were established. A Central Research Department was set up, and also an Efficiency Department to concentrate on organisation and methods, time studies and similar pursuits. Major decisions were reached on production rationalisation: the closure of the older iron ore mines in Cumberland, the scrapping and sale of some redundant plant, the concentration of manufacture of rails on Workington, of other heavy rail products on SPT, of special steels on Samuel Fox. The committee system was recast and branch responsibilities clarified while head office control over capital spending, sales and current finance was drastically tightened. The firm's public relations were enhanced. Personnel policies also received major attention.[4]

In all this Robert Hilton was the moving spirit. Certainly, Walter Benton Jones helped by symbolising continuity, focusing boardroom deliberation, building up morale and, in addition, overseeing colliery matters. But by and large, it was Hilton's ideas and experience which lay behind the new formal schemes, Hilton's contacts and interviews which decided the new managerial appointments, Hilton's compelling authority which breathed dynamism into the managerial system. Fortunately, Hilton's letter books, starting in July 1929, have survived. Some 120 of his letters, covering the latter part of 1929 during this critical early period, are particularly revealing. Their subject matter embraces new appointments, long-range policy issues related to product developments, marketing, acquisitions and federation affairs, staff and labour matters, and, by far the largest category, current operational matters. Included in the latter was a stream of demands for explanations, mainly from branch managers, about short-period deviations, often in tiny sub-units, for instance related to faulty deliveries, customer complaints, under-production, excess coke consumption, short-weight or excess costs. Both Hilton's letters and the surviving committee papers confirm not only his personal ascendancy but also the compression of important decisions into this initial short period. They confirm, too, the peripheral role of outside consultants. A report by H. A. Brassert in September 1929, which basically approved existing policies, was used largely for public relations purposes. As Hilton put it: 'Brassert carries great weight in this country, in America and on the Continent...we hope to get in some new money as a result.'[5]

Robert Hilton's operational missives often reveal the brusque disciplinarian. For example, he wrote to a senior director: 'Please investigate promises of delivery from The Ickles as these do not appear to me to have been given serious consideration in the past', and, later, 'There seems to be an increased cost of production in most departments. I shall be glad if you would give me the reasons for this.' To the chief accountant he wrote: 'I see from the attendance book...that your man X is systematically late, also that Y does not sign in the book. Please rectify both points.' When a senior official sold scrap to some competitors, avoiding the chief purchasing agent, Hilton commented that he was sorry the offender was not around 'for his indiscretion to be brought home to him'. The chairman's brother, Charles Ward Jones, was sternly reproved when Rother Vale Collieries purchased plant direct, not through the chief purchasing agent. No difficulties had arisen with any other branch: 'Please follow the procedure with plant that you do with stores.' To Ward Jones, Hilton wrote caustically on another matter: 'I notice that these minutes all refer to spending money: were there no questions of policy discussed at the meeting?'[6] This brand of toughness was much to the fore during the early period when basic disciplines had to be imposed.

It was during the early years, too, that the main outlines of the USC's capital investment during the 1930s were established. Modernisation schemes at Workington, especially new coke ovens and blast furnaces; technical improvements at Appleby and Frodingham; further forward diversification in Sheffield; the development of special steels at Samuel Fox: these guidelines were already clear by 1929. Even the major investments of the late

1930s were within the spirit of the early plans, notably in electric steel-making and stainless sheet plant at Fox, an ore-stocking scheme and new iron ore mine in Cumberland, a big expansion at Appleby-Frodingham, and the joint project on strip with John Summers. Table 7.1 shows the branch allocations relative to their profits. It demonstrates the emphasis on Workington and West Cumberland, which entailed relative sacrifices for the Sheffield units (for the complex motivations here see Chapter 8).[7]

Table  7.1    *USC: Branch Capital Expenditures and Profits 1931–9*

|  | Total capital expenditure (£,000s) | % | Total profits (£,000s) | % |
|---|---|---|---|---|
| SPT[1] | 478 | 5 | 3284 | 33 |
| Samuel Fox | 852 | 10 | 314 | 3 |
| Workington | 1281 | 15 | 657 | 7 |
| Rest of Cumberland | 54 | 1 | 14 | – |
| Appleby-Frodingham | 5109 | 58 | 4265 | 44 |
| Rest of combine[2] | 956 | 11 | 1281 | 13 |
| TOTALS | 8730 | 100 | 9815 | 100 |

*Note*: (1)   Including United Strip and Bar
       (2)   Mainly Rother Vale Collieries
*Source*: Andrews and Brunner, *Capital Development in Steel* (1951), pp. 205, 207

Overall, the policy was not what might be described as maximum profit seeking, although improved profits were obviously, and for some time crucially, a cornerstone. Thus enormous additional resources were to be concentrated on West Cumberland, even though little profit was expected therefrom. Thus the fact that the Rotherham works had higher costs than alternatives in Lincolnshire or Northamptonshire was not seen as proving that Rotherham was 'uneconomic' or that resources should be switched from it. Thus, again, it was repeatedly assumed that the search for new outlets was preferable to cuts so that the latter were not even fully investigated. A similar emphasis was apparent in thinking about internal controls. For example, the policy on intra-combine resource transfers emphasised conservation, mutual help and overall self-sufficiency rather than a profit-maximising search which might dictate external sales or purchases of raw materials with a sharp eye on market trends. As Benton Jones put it in 1930: 'When markets will not absorb our raw materials, the Combine *must* use them [my italics] to the exclusion of others, otherwise all the advantages of Combination would be lost.' The internal allocation of orders was increasingly decided on a standard cost basis, a yardstick closer to overall maximising than internal branch profitability but still removed from a systematic optimising calculus. There was a consistent preference for overall system maintenance in which the stronger sub-units were to help the weaker ones.[8]

The central idea was to increase productivity. This was partly a correlate of the aversion to both massive cuts and massive growth. For if efficiency could be sought neither by cutting out the weak nor by an aggressive expansion which would spread overheads and add new profit-makers, then an improvement in input : output ratios was needed over an extremely wide front. Partly because of Hilton's engineering background, partly under the auspices of 'scientific management', productivity was also viewed as a compelling objective in itself and one with enormous scope. Hilton's formulations on productivity were characteristically specific. For example, 'Fuel Economy, Time Studies...traffic problems and other questions all of which make for efficiency and which the average Works Manager whose great idea is output, is inclined to pass by'; 'better control of stocks and a more co-ordinated policy between the Works side and the sales'; 'Our greatest need is to reduce the production cost of our materials'. Correspondingly, he was contemptuous of vague and merely fashionable notions: 'The blessed word "Rationalization" is not fully understood in the Industry itself, and certainly not in Financial or Government circles in London.'

A notable feature of the USC's first efficiency drive was the importation of techniques not only from overseas iron and steel industries but also from other British industries both new and old. Thus an interest in electric furnace costs in New York or in the German and American 'plans' on fuel economy marched hand in hand with much borrowing of ideas from the electrical engineering industry, especially Metropolitan-Vickers, on research, sales techniques and apprenticeship schemes. It co-existed with the idea of cross-fertilising certain types of administrative skills. For example, a senior retired man from the railways was seen as the ideal person for 'the perfecting of the detail organisation' at Templeborough (increasingly restricted to large, standard manufacturing operations), even though 'he does not know anything of steel making' and his appointment 'raised certain criticisms'. Some of the management technology transfer was secured by overseas visits, as when Hilton proudly reported that a young manager had returned 'a changed man' from a study of German organisation and methods. Mostly, though, it came through an influx of new managers, particularly into sales and the newly created central efficiency units.[9]

Productivity-raising went beyond a string of techniques to embrace a larger concept of managerial orchestration. Better uses of labour, money and materials could not be secured without well-rounded, highly motivated junior and middle managers whose own productivity was improving. Here Hilton's interest in the idea of well-educated (but not *over*-educated) managers, his criticism of university remoteness from industry and of parental snobbery about 'clean collar jobs' provided impetus. So did his concept of the ideal executive, balancing 'personality, reasonable intelligence and a sense of fairness in dealing with men'. Again, co-ordination required improvements in horizontal communication between managers in different branches and functions, in consultation (so the lower échelons could contribute to and understand something of the rationale of corporate policy) and in 'general managership' (whereby much overall co-ordination of functions would be pushed downwards to the branches). On these fronts progress was sought zealously, if not always successfully. A management apprenticeship scheme was introduced in 1929 whereby a number of promising entrants from universities and from public and grammar schools were recruited each year and given some two years of varied grassroots training. Hilton paid considerable attention to the promotion and development of top management, including cross-postings and deliberate broadening of experience. Benton Jones cultivated morale with branch visits and speeches to employees. Company-wide conferences of managers were held, starting with one for sales staff in 1929.[10]

Meantime the top management was shifting towards youth and professionalism. The board lost its two elder statesmen with the death of the revered Mannaberg in 1930 and the retirement of Sir Frederick Jones. The conservative veterans J. V. Ellis and Francis Scott Smith had to be eased out, the former resigning in 1928, probably under pressure, the latter having his roles circumscribed by degrees before he, too, resigned in 1931. Of the remaining old guard Albert Peech, the ex-chairman, stayed out of the way, concentrating on railway business in London, and James Henderson continued for several years as general manager at Appleby-Frodingham. Some glaring gaps on the board were filled at last. J. Ivan Spens (appointed in 1928) brought much needed expertise on external finance. Robert Crichton (appointed in 1929) became a very active director as well as a badly needed dynamic and morale-boosting force as general manager at Workington.

An enthusiast for 'new blood', Hilton curtailed the dynastic element but probably less than he would have liked. Several scions of the USC's leading families he respected as capable: Gerald Steel, whom he sent to establish an Indian branch and later put in charge of Samuel Fox, Ronald Steel, Gerald's brother, and Langton Highton, groomed as a potential Workington general manager. Among those who fell short of Hilton's high standards, a young Tozer was quickly sent packing but two older dynasts remained on the board, J. E. Peech, who was confined to minor duties, and the chairman's idiosyncratic brother, Charles Ward Jones. Hilton's curtailment of the younger family elements caused resentments, some quite unjustified.[11]

Much of this activity represented no more than a process of catching-up with advanced practices elsewhere. This was probably particularly true of information systems. By 1928–30, for example, it was hardly distinguished pioneering to develop basic statistics on product end-uses or to provide for depreciation, interest and overheads in estimates of profits or losses 'in order that the results should not be misleading'. In its approaches to marginal costs and transfer pricing the USC probably continued to trail behind Stewarts and Lloyds. For instance, central decisions on both internal transfer prices and external prices at Templeborough still showed some confused thinking in 1930. Where the USC genuinely set the pace was in the speed of its overall achievement, in the sophistication of some of its techniques and, above all, in its fruitful testing-out of the problematic frontiers between 'centralisation' and 'decentralisation'.

Speed brought not only a classic turn-round but also the outlines of a whole new structure within about four years. Sophistication was shown over a wide front. In addition to organisation and methods, work study and management training, the systematic evaluation of investment projects was established (starting in a small way in 1928, with two-year capital forecasts and budgets added in 1930, three-year ones by 1931). Overall budgeting and planning for one year ahead, including forecasts, standards and analyses of variances *vis-à-vis* costs, output, sales and profits were operating, if crudely, by 1930. Successive improvements followed, including three-year profit targets by 1936.[12] But it was in grappling with the 'centralisation/decentralisation' issues that the USC met its biggest test.

## Centralisation versus decentralisation

It is obvious that the combine was becoming more centralised than before and deliberately so, particularly during the first period. For this there was much leeway and arguably an urgent need. But the degree and type of centralisation achieved in practice might differ from formal objectives, themselves hard to define. Thus H. A. Brassert eulogised the USC's policy of 'centralising sales policy and general administration, but of decentralising in respect to works operations'. This, however, reflected, perhaps even accentuated, some confusion over both aims and realities. All the more so when Brassert went on to praise the *'esprit de corps* which is the invariable result of placing full responsibility on departmental heads, backed by the fullest support from the head office'.[13] After all, was it likely that 'policy', whether for sales or anything else, could be completely centralised? Could 'full responsibility' be other than split? Was head office's main function lower down reducible to mere 'support'?

In practice, the headquarters under Hilton took firm control of the commanding heights of capital expenditure and finance, managerial appointments and general strategy (overall planning, mergers, acquisitions, etc.) while the branches had control of current operations. But even apart from definitional problems on these terms, there were vast grey areas between, reflecting a number of largely informal currents which acted independently or even in defiance of the official theory. Some of these currents produced more centralisation than was admitted. Thus the effective power of the central service departments often exceeded their formal remits. Thus, too, a powerful, often astringent, chief executive like Robert Hilton might create, sometimes unintentionally, lower-level anxieties and subtle pressures to conform. But some other currents worked informally towards decentralisation as when considerations of morale, communications and the need to develop managerial initiative suggested deference to lower-level units. Certainly, some formidable influences made for *divided* responsibilities. These existed in the generation of forecasts, targets and plans over a wide front (see below). They reflected the fact that the board included some heads of branches, which created a definitional overlap and enhanced their power. Divided central–local responsibilities also existed in the sales field. Here head office control of a large, dispersed, general sales force was justified by economies of scale. But in most branches the responsibility for sales policy appears to have been defensibly shared between the branch general managers and the top management.[14]

The problem of defining, let alone achieving, an efficient balance between centralisation and decentralisation was thrown into sharp relief by two issues. First, there was the role of the head office functional departments. Hilton formally adhered to orthodox organisational theory to the effect that these should 'service', 'advise' and 'help' rather than dictate to branch managements. But even formal definitions of the units' roles sometimes slipped into semantic confusion. This happened, for example, when Hilton wrote of a new chief accountant being able to 'direct the chief executives at the various branches' or when he explained, on the one hand, that the new efficiency department 'exists merely to make suggestions to management and takes no credit for such suggestions' and, on the other, that through its agency 'HQ is *only* co-ordinating the efforts of the various Works' (my italics). The efficiency department's very name was an irritant, as Hilton himself recognised. Its officials went out to branches to study operations and were widely regarded as head-office agents and tale-bearers. Much of their work was intended to correct the faults of older, traditionally-minded works managers who were highly sensitive. It naturally took time for these conflicts to erode through processes of education and assimilation, the retirement of the old men or the absorption of the efficiency experts into the line management.[15]

By the late 1930s a new wave of efficiency promotion was proceeding, this time focused by the accountants. In 1935 Hilton set up a committee under H. A. Simpson, comptroller and chief accountant, with the object of unifying and improving methods throughout the USC, particularly budgeting, financial and works accountancy, clerical organisation, and works progress and planning, extending by degrees to such matters as transport logistics, raw material control and labour turnover statistics. If anything, this second effort caused more controversy than the first. Again, it involved a head office network throughout the branches, this time largely composed of accountants already *in situ*. Again, this probably was, and certainly appeared to be, more coercively centralising than the official formulas suggested. But the resulting controversies also reflected personal and structural factors. Simpson, a man who brought an almost religious zeal to the cause of efficiency, had a sharp, edgy, autocratic personality. He also enjoyed Hilton's protection, another factor which sometimes made for fear and suspicion. Moreover, the aim of defining standards for 'each detail of every department and service in agreement with the departmental managers', extending to the expected lives of chipping hammers and grinding wheels, risked an obsession with trivia and with tidiness for its own sake, dangers which an accounting bias was perhaps unlikely to correct.[16]

By this time there were psychological obstacles to squeezing higher performance out of a system which had been through a vast overhaul and which could now, in many people's minds, rely on general economic prosperity to see it through. A shrewd contemporary observer noted an inverse relationship between the prosperity of a branch and its amenability to the new Hilton–Simpson disciplines: at one extreme Appleby-Frodingham, well-heeled and resistant; at the other Workington, indigent and obedient.[17] For the USC management to mount such an effort in the late 1930s was, perhaps, the greatest tribute to an efficiency orientation which reflected not just fears of another recession (particularly marked in Benton Jones's case) but also underlying concepts and habits of mind. If that effort squeezed out further advances, it also highlighted the way in which efficiency promotion in a large business, especially a prosperous one, necessitated more centralisation than was admitted and much that was disliked.

The other centralisation versus decentralisation problem was that of responsibilities for planning. These were almost inevitably joint and so often confused. An elaborate planning–budgeting system required extensive branch participation. The collection of large amounts of data from sub-units would be impossible unless the branches appreciated the underlying rationale. The generation of forecasts, notably of sales, called for careful, creative thinking at junior levels. Ideally, too, policy thinking required inputs of ideas and suggestions from below. Yet these requirements complicated the demarcations. If extensive consultation took place about general policies, as happened in 1931, participation and morale as well as thoroughness might be served, but not speed. Information could sometimes be obtained only by an informal threat system applied by head office officials. Top management repeatedly complained about the poor quality or biases of sub-unit forecasts: sloppy sales budgeting delegated to mere clerks; capital estimates which were 'merely jumped figures' or 'a covering figure in arriving at which everybody is playing for safety'; over cautious forecasts; 'a wide variation between Actual and Budget results which called for more care on sales budgets and closer contact between accountants and both production and sales'; trading estimates which were 'defeating the whole object of budgeting' and obfuscating 'the formation of policy'.[18]

Correcting such problems involved head office either in education, a long-term process, or else in close immediate surveillance, which again diluted delegation. Overall, it is likely that the USC's planning–budgeting system both reflected and encouraged a continuous interaction between head office and the main sub-units, a process in which the former was more powerful but the latter were still important. Whatever else this was, it was a far cry from tidy notions of centralised *diktat*, profit-centre autonomy or even a clear demarcation between the two.

Through all this experimentation the outlines of an enduring organisational achievement can be glimpsed. Behind it lay a great improvement in morale which was partly, of course, based on transient factors. There was the sheer exhilaration of the 1928–32 turnround. There were the potent mixed incentives provided by the Benton Jones-Hilton duumvirate with its blend of monarchy and premiership, charisma and disciplinarianism, although the latter element, with its accompanying astringencies, was to recede after Hilton's retirement in 1939. But good morale also reflected a more permanent corporate ethos with which many sections could identify morally and emotionally. That ethos involved behaviour consonant with many of the public norms of the 1930s, particularly on employment maintenance, labour policy and industrial co-operation (see Chapters 8 and 9). The resulting feeling of being a 'decent', 'humane' and 'public spirited' organisation was probably a widespread integrative force. So was the broad way in which 'efficiency' was promoted as a dominant corporate objective: efficiency in developing human resources as well as in improving quantifiable productivity, efficiency in the longer-term as well as the short.

This ethos, in turn, was one reason why the USC pursued, however imperfectly, a balance between centralisation and decentralisation. The main enduring features of the balance were (1) centrally monitored information systems and productivity promotions, (2) participative planning–budgeting processes, and (3) a multi-divisional system of a few large, mostly multi-product branches. Because of the scope provided by (3), branch managers could become true general managers, co-ordinating the different functions and activities at local levels. It was easier, therefore, for a new generation of effective all-round managers to develop. The branches focused on well defined locations, each with its own traditions and distinctive character, so the 'spirit of place' could be canalised in the service of employee morale. Moreover, each branch employed thousands rather than tens of thousands, so that some of the evils of excessive size in human terms could be avoided. Overall, this pattern reflected a holistic view of efficiency and it implied a continuous search for some 'golden mean', an iterative exploration between the extremes of centralisation and decentralisation. In these ways it is arguable that by 1939 the USC had struggled through, as much by accident as design and certainly with pain and error, to a pattern of organisational effort that was efficiency-promoting in a balanced way, sustainable over a long period and a useful example to others in the industry.

## Ultra-centralisation, belated rejuvenation

Meantime the other two firms were not standing still in the field of efficiency but their movements significantly diverged from the USC's. Despite some overlaps the differences were striking. Relative to the USC, there were deficiencies of breadth and balance in one case, of timeliness in the other. Broadly speaking, Stewarts and Lloyds' achievement was narrower while Dorman Long's was yet again delayed. Both of these salient contrasts can be reviewed quite briefly.

In Stewarts and Lloyds, as we have seen, organisational efficiency had been present as a dominant cause even before 1914 (see Chapter 3). That cause had been furthered over a wide front in the early 1920s (Chapter 4) and continued to receive a strong emphasis after Allan Macdiarmid's succession in 1926 (Chapter 5). Moreover, the firm's efficiency pursuit had followed a consistent pattern of centralisation in the cause of 'rationalisation'. Basically, what happened in the 1930s was an acceleration and intensification of this trend. Although the resulting organisational pattern accorded quite well with Stewarts and Lloyds' central growth strategy, discussed in Chapter 6, it also had some disadvantages.

Further centralisation had much to do with the play of personalities in the boardroom from 1932 onwards. Just below Allan Macdiarmid were two key men, both of them managing directors: Guy McClay, a brilliant engineer and ideas man particularly in tubes, rough mannered, a backroom advisory figure, and something of a 'loner', and Francis McClure, a tough, subtle, hard-driving salesman and top negotiator. If McClay made up for Macdiarmid's technical limitations, McClure reinforced him commercially. Both had strong ties with the chairman. McClay was very much Macdiarmid's man, having followed him upwards in the firm. McClure was also his double brother-in-law and a close friend. When Macdiarmid moved from Glasgow to rural Hertfordshire in 1931 both McClay and McClure went too, moving to houses within a few miles and forming, with their families, a small coterie of Scots expatriates. The three men frequently met out of office hours, mostly at Macdiarmid's home, in an atmosphere that blended the house party and the boardroom cabal. Another close neighbour, attached warmly, if more loosely to the circle, was the influential consultant, H. A. Brassert. Of course, there were other leading figures, notably the joint managing directors, Joseph Howard, J. H. Lloyd and A. G. Stewart (the previous chairman's son and a rising star); somewhat later, R. Menzies Wilson, a tough administrator who became operating chief at Corby; and Nigel Campbell, a City man with important contacts, who advised on external finance and proved politically useful. But it was the central trio, essentially a giant reinforced by two *alter egos*, which lay at the heart of the centralising drive.[19]

As to the intensity of that drive there can be little doubt. The concentration on Corby facilitated still tighter controls over finance, purchasing, internal transfers and overheads. Both the sales force and the sales policy were brought under the direct control of the sales managing directors, McClure and Lloyd. Unified command of coal, iron and steel had operated from the early 1920s only over the newer sub-units, while in tubes generally it had been long delayed. Now at last the control of production as a whole was centralised, helped by a new central research department, formed in 1929, which was transferred to Corby in 1934 and re-christened the department of research and technical development.[20] All this went alongside the generalist and functional composition of the board, a longstanding policy, and the deliberate attempt to downgrade geographical and product separatisms at that level. From 1932 there was a drastic consolidation of committees into a single top managing committee of the five managing directors plus Macdiarmid as chairman and general managing director. There was Macdiarmid's pre-eminence as effectively both king and prime minister (whereas in the USC, for example, Benton Jones and Hilton divided these roles), further reinforced by the strength of the Hertfordshire inner circle. For instance, it was Macdiarmid, not the managing committee, who made reports to the directors, and these reports were not even pre-circulated, making the board still more of a rubber stamp. As to the managing committee itself, a later internal document was probably not far wrong when it observed wryly: 'As was made plain then and on subsequent occasions, the word "committee" was really a misnomer as the members were in effect advisory to the G.M.D. [Macdiarmid]'.[21]

Of course, Stewarts and Lloyds' high-pitch centralisation had some corporate advantages. Theoretically, in Chandlerian terms, a centralised, functional organisation was consistent with a high degree of product homogeneity and specialisation. It also married well with the atmosphere of secrecy and almost military precision which attended both Stewarts and Lloyds' international role and its domestic invasion of iron and steel. On the other hand, there were disadvantages. The strain imposed on a few top men was recognised internally by 1941.[22] The system probably constrained managerial initiative lower down and tended to develop senior people who were specialists rather than broad general managers. It did not encourage participation, since planning processes were restricted almost solely to the boardroom, and from that viewpoint it cannot have helped morale (although it should be added that other factors, including the firm's outstanding external successes, were positive influences on the latter). The implied view of efficiency was one-dimensional. Rationalisation and economies of scale across the familiar spectrum of financial flows, purchases, internal transfers, production and sales continued to monopolise attention. Ideas of scale *dis*economies, human resource development and communal morale-building hardly featured.

To some extent Stewarts and Lloyds was paying the price of its earlier leadership on efficiency issues. It was consolidating and refining familiar ideas rather than breaking new ground. The fact that ultra-centralisation coincided with a striking coming-into-effect of economies of scale as a result of Corby helped to mask limitations which would, in any case, take some time to emerge. It is arguable that these limitations were, in fact, severe according to the criteria of management development, sustainable organisation, ability to handle diversification strategies and long-term morale. Although still marked, therefore, the firm's efficiency-seeking in the 1930s lacked something of the sweep and breadth of the USC's.

Meantime, Dorman Long's main contrast with *both* the other firms lay in the fact that its move towards an efficiency phase was so long delayed. Once again here was a case of arrested development or, to be more precise, arrested rejuvenation. In fact, the long saga of managerial woes chronicled in Chapters 4, 5 and 6 did not close until the end of our period. Only in the late 1930s, following a further prolongation of the firm's Golgotha, were there glimpses of some sort of resurrection.

In terms of efficiency, the Mitchell years (1931–4) brought an improvement in current financial controls, but little else. Just to take one example, outsiders were startled by evidence of blatant competition between the old Dorman Long and the newly 'amalgamated' Bolckow Vaughan constructional interests as late as 1933. The legal contests over the South Durham merger scheme in summer 1933 led to a confused inter-regnum of several months. Dorman Long and the BID Co hoped to revive the merger scheme, although the South Durham leadership was evasive and privately scathing: in Talbot's words, Dorman's 'financial mess' would have to be cleaned up first. Then the final court judgements produced further humiliation. In order to avoid a Receivership, Dorman Long was forced to bend the knee to its debenture holders. By January 1934 a powerful debenture holders' committee, parleying with both the company's board and its bankers, began to dictate terms for 'getting rid of "white elephants" and strengthening the management of Dorman's'. Its chairman, Sir Miles Mattinson, was a trenchant critic of 'bad trade in the past...unfortunate commitments...extravagance...excess of capitalisation on amalgamations which had not prospered...splendid bridges which did not pay'.[23]

One result of the Mattinson Committee's dominance was predictable. In March 1934 Charles Mitchell resigned from both the chairmanship and the board. Mitchell had always been somewhat isolated inside Dorman Long. His London concentration had caused resentment in Middlesbrough. The discovery in early 1933 of a fraud perpetrated on the company by its chief accountant over several years had left an impression of personal negligence. The merger débâcle had not helped Mitchell's reputation either. Now his departure probably seemed the best tangible concession to a debenture holders' committee breathing fire and avid for sacrifices. Mitchell was given £5,000 compensation and the press announcement emphasised administrative changes as the reason for his going. As already emphasised, his was not the main responsibility for Dorman Long's basic weaknesses and long accumulated woes. But it was a pity from every point of view that he had not left earlier.[24]

The way was now clear for one of the most remarkable managerial takeovers in the history of interwar steel, the slow, behind-the-scenes assumption of power by Ellis Hunter (1892–1962).[25] Hunter, a Peat, Marwick, Mitchell senior accountant, had audited Dorman Long's accounts and gained an intimate knowledge of its financial convolutions. He had won the confidence of the Dorman Long board, also of the debenture holders' committee whose trusted adviser he became. Through 1934 Hunter's negotiating skills, knowledge and detached demeanour made him an increasingly influential intermediary between the committee, Barclays, and Dorman Long's now somewhat disorientated board. It was Hunter who probably inspired various changes after Mitchell's departure: sundry retirements of senior officials, draconian cuts in the London office staff, the elimination of the roads department, salary reductions, a consultant's investigation of the firm's collieries, a quagmire which the Bolckow Vaughan merger had deepened. More two-edged, perhaps, was the return to leading roles of Maurice Bell and W. L. H. Johnson, the family appointees previously demoted by Mitchell. It was probably this which led to an impression inside the BID Co that 'they [Dorman Long] are getting "the old gang" back'. But by October 1934 the remodelling appeared far-reaching. Laurence Ennis, the most senior non-family executive director, had been made managing director. The board was reconstituted, with no less than five resignations, and a new outside chairman was appointed, the well-known politician and City figure, Lord (Harmar) Greenwood: an obvious bid to restore the firm's poor public and financial standing. Ennis, however, a long-time specialist and now elderly, could not provide vigorous overall leadership, and Greenwood was not (and was probably not intended to be) anything more than an imposing figurehead.[26]

The major gap that remained was to be filled by Ellis Hunter. The son of a North Yorkshire schoolmaster, educated in Middlesbrough, Hunter was an intensely shy man, withdrawn, reclusive, forbidding in manner, sometimes socially awkward, and his background had been purely in finance. But in addition to the intellectual and negotiating abilities just mentioned, he possessed great advantages as a chief executive for Dorman Long at this point: decisiveness, a blend of the 'new broom' with inside knowledge, a tall, craggy, imposing presence partly reminiscent of old Sir Arthur Dorman, a 'local boy made good' image that was also beneficial to the firm's much tattered morale. Hunter's reputation waxed as Dorman Long benefited from the post-1935 economic recovery. The years of his ascent were marked by a (mainly externally derived) restoration of profits, a big increase in employment, an excision of the old cancer of over-capitalisation, and a surge of capital investment, notably in a programme of new coke ovens.

Judgement of Hunter's efficiency promotion is difficult partly because so much of it lay ahead, partly because even the best of company reforms knocked up against certain historical constraints. For example, Dorman Long still had four geographically separated works, each with blast furnaces supplying hot metal in ladles to their steel making furnaces, and harbour and wharfing facilities remained inadequate. Nonetheless some clear lines of achievement were emerging before 1939. They included rigorously conservative finance, a clean-up of the subsidiaries, a new Central Engineering Department with great potential, and a decisive move away from nepotism. This programme lacked many efficiency-promoting techniques adopted by others, for instance standard costing, budgeting–planning and formal management training. It was very much a matter of catching up on elementals, making sure that people did things precisely, conscientiously and on time, restoring minimal disciplines: tasks for which Ellis Hunter, a supremely schoolmasterly figure, was, it seems, well equipped. There can be no doubt that Dorman Long had at long last found a managerial saviour and that a phase of concentration on efficiency had begun.[27]

## Conclusions on growth and efficiency

Having considered the 1930s experience on growth in the last chapter and on efficiency in this one, we are now in a position to relate our findings to the evidence from the earlier chapters. It should be possible to reach some tentative conclusions at least on those parts of our theory which relate to growth and efficiency (social action will be considered in the next two chapters). How far does the overall experience of the firms from 1914 up to 1939 support the idea that there are long-term, systematic relationships between growth and efficiency pursuits as a reflection of corporate and managerial forces? More specifically, does this experience support the three basic propositions? These are:

1  Managerial phases related to growth (G) and/or efficiency (E), sometimes interspersed with periods of indeterminacy (I).
2  A long-run sequencing of these phases along the lines of reiteration (G+ E → G + E ... ) or alternation (G → E → G → E ... ).
3  A tendency for the phases to be biased (and hence for alternations to dominate over reiterations in the long-run) in so far as personal power concentrations, time constraints and managerial specialisations combine.

With regard to (1) and (2) the pattern can be summarised thus. Chapters 3 and 4 suggested that in Dorman Long the dominant trend initially was growth or growth-seeking. Now armed with the further evidence about the 1930s, we can see how attempts to continue growthmanship even further contributed to a slide into indeterminacy from 1931 until the mid-1930s, by which time a belated efficiency phase began. So the Dorman Long sequence emerges as G → I → E. Similarly combining the evidence on the USC from earlier chapters with that for the 1930s produces a comparable sequence G → I → E. In the USC, though, the timing was different since the points of transition were 1920 and 1928. In the case of Stewarts and Lloyds the pattern emerges as G + E →

G + E, albeit with some difference of emphasis and relative success as between the periods up to and after 1926.

Of course, certain differences have been observed within these broad behavioural tendencies. There were marked variations in personality and management style so that, for example, Harry Steel's growthmanship had a different flavour from Arthur Dorman's. The character of the changeovers significantly varied. It should also be emphasised that the 'growth' and 'efficiency' tendencies comprehended important differences of emphasis, for instance as between acquisition and investment-led growth in the former case, and production and wider organisational efficiencies in the latter. Again, the USC's indeterminacies in the early 1920s were more two-edged than those of Dorman Long by the mid-1930s. For that matter, a similar phase in the *same* firm could have different connotations so that the Macdiarmid regime's brand of growth plus efficiency was more intense than its predecessor's. Nonetheless, despite these qualifications, the broad characterisation of the phases is reasonably clear. Evidence from a wide variety of sources has been presented in support of the basic distinction between the four types of situation envisaged in (1), namely G, E, G + E or I. Moreover, it seems valid to characterise such phases as 'managerial' in that they broadly followed the timing of successive managerial régimes. The eight managerial régimes studied seem to fall fairly clearly within the suggested typology.

With regard to proposition (2) the time scale of observation is restricted and the limitations of a tiny sample are particularly acute. There can be little doubt that Stewarts and Lloyds' success in combining growth and efficiency up to 1926, if only at moderate levels, produced a strong internal desire to repeat that blend, if possible at a higher pitch. Equally, there can be little doubt as to the role of dialectical corporate reactions against previous excesses in the other two firms.

That Harry Steel's 1916–20 empire-building created a backlash reaction within the USC is shown by the 1920–7 régime's nervousness about further mergers and by the feeling of its ablest senior men, notably Mannaberg and Sir Frederick Jones, that the combine needed not further expansive adventures but a rationalising consolidation. Even without the depression some boardroom scepticism as to the potential for still further elephantine growth, together with desires for proper absorption of existing resources, would probably have developed. The fact that an efficiency reaction only became effective after 1928 merely shows how a suspension of strong managerial pursuits can delay the working-through of a dialectical process which demands changes at the top. In Dorman Long, too, the long-protracted growth-manship until 1931 generated internal reactions: a distaste for further adventures, a desire for greater efficiency. At least by the late 1920s, the view that things urgently needed to be tidied up was held by Charles Mitchell, the Chief Accountant, W. H. Davies, and probably also by Laurence Ennis. The Bolckow Vaughan merger, itself no immediate contributor to efficiency, brought in several senior men who thought likewise. There can be no doubt that this backlash contributed to the partial efficiency measures of 1931–4 and the more decisive swing towards efficiency in the late 1930s. But here again the internal reaction effect could only become effective with a decisive change of managerial régime. The interwar experience therefore supports the idea of corporate processes being partly responsible for a reiteration (in Stewarts and Lloyds) and for two cases of (interrupted) alternation of the G → E type. On the other hand, no cases of an E → G alternation emerge.

With regard to proposition (3), biases emerge in the case of six leading figures: Dorman, Steel, Mitchell, Benton Jones, Hilton and Hunter, in the first three towards growth, in the last three towards efficiency. To repeat, this does not mean that these men were uninterested in the other pursuit or unaware of its importance, but simply that their values and abilities pulled them naturally towards a particular bias. The fact that this individual bias could often be highly creative and dynamic, all the more so because of its intensity, is another point. Only in Macdiarmid's case is there clear evidence of growth and efficiency pursuits resonating with some equality inside an individual. Moreover, the biases appear to have been deep-seated and prolonged, both pre-dating and persisting throughout the periods of top managerial tenure.

As has been frequently emphasised, such managerial specialisation would produce *corporate* performance biases only in so far as power was concentrated and time constraints severe. With regard to the former, the evidence suggests a strong overall tendency towards concentration of power on two or three individuals, usually with just one pre-eminent. The cases of individual dominance (Dorman, Steel, Macdiarmid) and dual dominance (Hilton–Benton Jones) outnumber those of control shared among a few (the J. G. Stewart, Peech and Mitchell régimes). As for the time constraints, the accidents of geography were at work. In Stewarts and Lloyds the possibility of uniting a London-concentrated growth strategy with surveillance of grass-roots operations was helped by the increasing Corby concentration and by the fact that from 1931 the Hertfordshire inner circle was based rather neatly within easy distance of both London and Corby. By contrast, the decidedly provincial concentration of both Dorman Long and the USC made it hard to reconcile proper operational control with spending a lot of time in London. In the USC, top management's natural concentration on its main Sheffield–Scunthorpe–Workington axis probably harmed its access to power and growth centres in London (where Albert Peech was a poor substitute). In Dorman Long, Charles Mitchell largely operated in London between 1931 and 1934 to the apparent detriment of efficiency pursuits on Teesside. Not until the late 1930s, with the combination of a public-relations and politically-oriented chairman in London and an operational chief executive in Middlesbrough, was this problem on its way towards resolution.

A bigger time constraint lay in the fact that in both firms a reaction towards efficiency would consume enormous resources of managerial energy. This was particularly so when the new leaders were also newcomers to the industry and/or the firm. If such freshness helped the psychology of the 'new broom', it also necessitated spending much time on learning about the organisation. This was the case with Hilton in 1928–9 and, to a lesser extent, Ellis Hunter in the late 1930s. But in any case a serious reformation of both the USC and Dorman Long required a lion's share of top management time and energy, drastically cutting down what was available for growth pursuits.

But although power concentrations and time constraints could intensify the biases, we still come back to the basic factor that these men gravitated naturally towards what they believed in and what they were good at. Thus in the USC both Hilton and Benton Jones tended towards efficiency rather than growth as a matter of temperament and ideals, talent and experience. Hilton was no empire-builder or power strategist, no consummate lobbyist or subtle negotiator but a disciplinarian, a reformer and a brilliant administrator. Benton Jones lacked the power urges, the idealism of creative growth, the competitive aggressiveness, the streak of killer instinct, the wheeling-and-dealing abilities associated with strategic growth. To him ideas of conservation and order, method, equity, human resource fulfilment came more naturally. Again, Charles Mitchell was theoretically attached to both efficiency and growth. But he had long concentrated on constructional business and his apparent strengths were as a salesman and wheeler-dealer. Whether this fitted him by sympathy or competence for efficiency promotion is doubtful. It was his misfortune, as a likelier candidate for growthmanship, to inherit a giant expansion project which was doomed to failure and which cheated him of success even there. Conversely and more clearly, Ellis Hunter's background in auditing and financial consultancy contributed to a strong efficiency orientation. Again, his values and strengths were in conducting and orchestration, not in composing a new work, in order rather than conquest.

There is some support, therefore, for the hypothesis of typical individual biases which concentrated on growth or efficiency and which powerfully influenced corporate phases of a similar type. But the theory of managerial specialisation is not restricted to these two facets. Nor can corporate performance and business policy be evaluated only in terms of growth and efficiency (or only in terms of profit). Some further facets and criteria also require interpretation, and it is to these that we now turn.

## Notes

1 Mainly U/DM and U/CC, also U/FJ and U/RA.
2 For the critical early decision, partially socially motivated, to maintain the West Cumberland operations, and also for the USC's relatively stable employment record generally, see chapter 8. For Benton Jones's aversion to cuts and divestments see particularly U/CI, notes 30.5.31, 22.6.31 and 15.7.31: for example, 'Is there any fundamental reason why this Branch [Workington] should not become profitable?'; 'The works [Rotherham] are there, they cannot be removed bodily and if we do not use them we must abandon them'. A similar view emerged with regard to the Samuel Fox Colliery, see U/CI, 22.6.31 and U/CC, 22.7.31. Only in 1931 did Hilton venture the tentative opinion, 'We may have to consider the shutting down of certain units', U/CC, 2.6.31. But the actual cuts were tiny.
3 U/DM, 4.4.29, 16.4.29, 14.5.29, 11.6.29, 9.7.29, 13.8.29, 8.10.29, 10.12.29, 25.3.30, 15.4.30, 20.6.30, 14.5.34, 4.10.34. U/MD, R. S. Hilton to G. Steel, 2.10.29 and 14.10.29, to R. A. Dyson, 4.10.34. Andrews and Brunner *Capital Development in Steel* (1951), pp. 156–62. BE: SMT 2/147, especially memoranda 29.5.29 and 18.9.29, and Montagu Norman, confidential memorandum, 21.9.29. Also BID 1/116 for 1934 reorganisation.
4 U/DM. U/CC. U/MD. Andrews and Brunner, op. cit., pp. 162–6.
5 U/MD. Interviews with retired USC officials. Andrews and Brunner interviews 1950, W. Benton Jones. R. Peddie, *The United Steel Companies Ltd* (1967), p. 19. Report on the Properties of USC by H. A. Brassert and Co., Sept. 1929 (006/10/1). U/MD, to G. Steel, 2.10.29.
6 U/MD, 14.8.29, 16.11.29, 26.11.29, 12.12.29, 15.2.30, 8.6.34.
7 U/DM. U/CC. Brassert Report, Sept. 1929, op. cit. For a detailed account of the USC's capital development from 1928 to 1939 see Andrews and Brunner, op. cit., pp. 166–74 and Chapter 6.
8 U/CI, notes 22.6.31 and generally in 1931. U/CC, 3.6.30, 22.10.30. U/DM, 9.9.30, 14.10.30. U/CI, Comptroller to Central Committee, 8.8.33. U/MD, to J. Henderson, 8.6.33.
9 U/MD, 20.8.29, 8.1.30, 8.8.30, 14.11.30. For the technique importations see U/MD, 13.7.29, 22.7.29, 20.8.29, 10.10.29, 22.11.29, 12.12.29, 22.2.30, 19.9.31, 28.6.35.

10   For general concepts see U/MD, to G. Steel, 6.8.30, 18.12.30 and 13.3.31, to
     Miss D. M. Hughes, 1.1.37; also U/CI, Notes, 22.6.31. For activities on man-
     agement development, training and communications see U/MD, to G. Steel,
     20.8.29, 14.10.29, 13.3.31 and 14.4.31, to R. Crichton, 28.1.31; Andrews and
     Brunner, op. cit., p. 166; USC General Matter file 1924–9 (006/10/3), Sales
     Conference, 31.10.29 – 2.11.29. Author's interviews with retired USC
     officials.
11   U/DM, 10.1.28, 14.2.28, 8.5.28, 14.8.28, 17.9.28, 12.3.29, 8.10.29, 7.1.30,
     12.8.30, 23.10.35, 27.5.36. U/MD, letters to Scott Smith, 1929 and 1930, to P.
     Lindsay, 6.8.29, 12.7.30 and 22.7.30, to H. Tozer, 10.10.29, to G. Steel, 6.8.30,
     to R. Crichton, 28.1.30, to J. E. Peech, 19.6.30, to J. I. Spens, 9.12.33. BE:
     SMT 2/63, 103/2. Andrews and Brunner, interviews, 1950. Author's inter-
     views with retired USC officials.
12   For the information system improvements see U/DM, 13.3.28, 7.1.30, 9.4.31;
     U/CC, 4.9.28, 12.9.30; U/CI, notes 30.5.31 and 22.1.32; U/MD, 31.7.29,
     20.8.29, 14.11.29, 25.1.30; and USC, General Matter file, 1924–9 (006/10/3),
     Sales Conference, 31.10.29–2.11.29. For the 1930 Templeborough pricing
     decisions see U/DM, 9.9.30, 4.10.30. For investment evaluations and forecasts
     see U/DM, 12.2.29; U/CC, 1.5.28, 8.8.28, 4.9.28, 19.9.28, 11.2.30. For the
     beginnings of budgeting–planning see U/DM, 8.7.30; U/CC, 12.9.30, 24.9.30;
     and U/MD, 3.11.31. For its later development see particularly USC Long-date
     budgets (152/11/4), Office Manager's Minutes, 1935–41 (152/11/2) and U/CI,
     notes and letters in late 1930s. Some of the problems of budgeting and
     planning are referred to later in this chapter.
13   Brassert Report 1929 (006/10/1), op. cit.
14   Mr T. S. Kilpatrick was particularly helpful to the author on this point. See
     also U/MD, R. S. Hilton to G. Steel, 13.3.31.
15   U/MD, 18.9.29, 15.10.29, 3.2.30, 14.4.31, 15.6.31. Author's interviews with
     retired USC officials.
16   USC, Office Manager's Minutes, 1935–41 (152/11/2) particularly correspond-
     ence, May 1935, 2.10.35, 7.1.36, 4.4.36, 5.6.36, various reports July 1936,
     6.1.37, 19.3.37, 6.11.37. Long-date budgets (152/11/4). Author's interviews
     with retired USC officials.
17   For this comment the author is indebted to the late Mr Philip Beynon.
18   For policy consultations see U/CI, notes and papers 1931, also Andrews and
     Brunner, op. cit., pp. 188–91. For head office squeezing of branches for
     statistical information see USC, Manufacturing Costs, SPT, S. Fox, Appleby-
     Frodingham, etc. (152/11/12), Comptroller to Secretaries, 19.7.32, Comptrol-
     ler to General Managers, 22.2.34, H. A. Bletcher to J. Monkhouse, 17.3.36.
     For head office complaints see U/CC, 12.9.30, 21.12.32, 8.1.36; U/MD, 22.3.30;
     Office Manager's Minutes (152/11/2), 2.7.36, 7.8.36, 4.3.37, 8.4.37.
19   S/CB for directors' tenures and standing committee memberships. S/BM,
     4.5.28, 26.7.28, 19.12.28, 8.3.29, 1.10.29, 25.10.29, 17.12.29, 2.9.30, 16.12.30,
     3.2.31, 10.3.31, 15.12.31, 2.2.32, 22.3.32, 12.4.32, 31.5.32. In early 1932 G. A.
     Mitchell resigned from the board and C. G. Atha left the company. Interviews
     with retired S and L officials and with Mrs G. Castle (née Macdiarmid) and
     Mrs E. Holderness (née Macdiarmid). Nigel Campbell, a director of Helbert
     Wagg and previously of SMT, and on good terms with Montagu Norman,
     became a staunch S and L protagonist. See BE: SMT 2/148, particularly
     Montagu Norman to N. Campbell, 5.10.32; correspondence on British Man-
     nesmann purchase in 1935; and Campbell to Norman, 18.7.39 on Stanton
     acquisition. Also SMT 2/149, Campbell–Norman correspondence in January
     1937 on S and L/USC conflicts.
20   S/GPC, 24.10.29, 29.6.31. S/MC, 10 and 11.2.32, 16.2.32, 7.12.32, 19.9.33. S/
     BM, 2.2.32. Steelworks Committee (1791/1/16), 12.10.31. Economy Commis-

sion (1791/1/21). 14.7.32, 29.8.32, 22.9.32. S/EC, Memorandum on Central Research Department, 2.4.29. Information kindly provided by Mr E.G. Saunders. Interviews with retired S and L officials.

21 S/BM, 15.12.31, 2.2.32. S/MC. History of the functioning and personnel of various committees, 1903–32, Secretary's Department, March 1954 (1791/1/17), 22.2.54.

22 S/BM, 25.2.41. See also 1954 (undated) memorandum, Chartridge Lodge to L. M. T. Castle in History of the functioning and personnel of various committees, op. cit.

23 D/FC. D/CI (1066/5/1), R. Niven to C. Mitchell, 18.1.33. Cargo Fleet (210(d)/15/18), Sales Conference 28.12.33. See also D/CI (1066/5/8), W. H. Davies to S. W. Rawson, 23.11.32. Interviews with retired DL officials. D/DM, 21.2.33, 19.9.33, 29.12.33, 23.1.34, 27.2.34, 20.3.34, 28.3.34, 17.4.34. D/FC, C. Mitchell to S. W. Rawson, 3.1.34. D/CM, including W. B. Peat to J. H. B. Forster 31.1.34. SD/DM, 5.10.33, 14.11.33, 12.12.33. SD and CF, Additional Materials on Merger, 1933 (234/1/33), B. Talbot to G. Sturt, 16.1.34. Cargo Fleet (210(d)/15/18), Sales Conference 28.12.33. BE: SMT 9/4, 18.12.33, 8.1.34, 15.1.34, 26.2.34; BID 1933/34; BID 1934/87, particularly note by N. L. Campbell, 8.2.34, meeting, 28.6.34, Dorman Long Moratorium, meetings, 12.7.34.

24 D/DM, 28.3.34. *Financial Times*, 29.3.34. *Northern Evening Echo*, 29.3.34. Interviews with retired DL officials.

25 The following characterisation of Hunter relies mainly on documentary evidence, interviews with retired DL officials and his local obituary, *Northern Echo*, 22.9.62. See also C. Wilson, A Man and His Times: a memoir of Sir Ellis Hunter (c.1968). D/DM, 20.3.34, 17.4.34, 18.5.34, 22.6.34, 20.7.34, 30.7.34, 21.9.34, 24.9.34, 9.10.34. D/FC, especially 18.5.34, 21.9.34. Office Administration Committee (1066/13/2), 30.4.34. At the request of the Debenture Holders' Committee, Hunter attended DL board and board committee meetings from early 1934: see Sidney Harbour Bridge Contract (1066/8/3), Secretary to E. Lawrence, 17.3.34. BE: SMT 9/4, 28.5.34, 4.6.34, 18.6.34, 2.7.34, 16.7.34. Information on Lord Greenwood from retired DL officials.

26 D/AGMs, 11.12.35, 9.12.36, 8.12.37, 7.12.38, 13.12.39. D/DM, 7.1.35, 31.3.35, 3.5.35, 17.6.35, 26.6.35, 13.9.35, 27.9.35, 8.12.36. Managing Director's Monthly Reports, 1936–38 (1066/16/11). Ellis Hunter became an executive director and deputy chairman in early 1938, see D/DM, 20.1.38. Laurence Ennis, managing director since 1935, had recently retired, see D/DM, 25.6.37.

27 For the constraints see BE: SMT 3/78, H. A. Brassert, Memorandum on the North East Coast and the Steel Industry, August, 1935; DL, Managing Director's Monthly Reports 1936–8 (1066/16/11), Private Wharves Report, 31.12.36; I.D.A.C. Papers, PRO/BT 10/27, Report of special committee of BISF on iron and steel industry dock and harbour facilities, CP/158/38; and D. L. Burn, *The Steel Industry 1939–59* (1961), pp. 85–6, 174, 250. For DL's rationalisation moves see D/DM, 31.5.35, 28.6.35, 26.7.35, 13.9.35, 17.1.36, 3.4.36, 1.5.36, 22.5.36, 26.6.36, 24.7.36, 30.10.36. Also D/CI, Bowesfield Steel Company (1066/5/10), Tees-side Bridge and Engineering (1066/5/11), Wade and Dorman (1066/5/11), Redpath Brown (1066/5/12); Managing Director's Monthly Reports, op. cit; Sidney Harbour Bridge Contract (1066/8/3), Memorandum on Bridge Department 7.4.37. Interviews with retired DL officials.

# 8 Labour and local community problems

As the last few chapters have shown, management's task in continuously reconciling growth and efficiency at a high pitch was already complicated enough and, in two of our three firms, unfulfilled. But it would be oversimplifying to see that reconciliation as its exclusive aim. It was suggested in Chapter 1 that managements sometimes pursue publicly-approved objectives which either posit a different concept of growth and efficiency than their own or which relate to other values. These further pursuits, characterised as 'social action', often conflict with the firm's own growth and/or efficiency, at least in the short- and medium-terms, and may relate only tenuously to these even in the long term. In Chapter 2 it was suggested that the interwar iron and steel industry faced these problems of social action particularly acutely, to some degree becoming a pilot for developments which have subsequently become more widespread.

It is convenient to divide the treatment of social action into two parts. The first, dealt with in this chapter, concerns decision-making *vis-à-vis* labour and local communities – a subject which cannot be pursued in great depth largely because of the relative lack of evidence. Consequently, only a broad indication of some contrasts between the three firms can be provided but one sufficient to suggest a major difference between Stewarts and Lloyds and the USC. The second aspect, on which much fuller data exist – relationships with public policies and government – will be covered in Chapter 9. As we proceed, it will become clearer that the balances between growth and efficiency discussed so far were themselves influenced by varying pursuits of social action, and that a fair overall judgement of the management of our three firms must also take these latter into account.

## General labour policies

In all three firms labour policies in the 1930s owed much to long-established attitudes and habits. Characteristic stances had developed over long periods, reflecting varying managerial values and styles. In Dorman Long, for example, Sir Arthur

Dorman's approach to labour matters was old-style paternalism, steeped in familial, even faintly monarchical ideas. Nostalgically, both he and Sir Hugh Bell invoked a golden tradition of close relationships felt to have been inherited from older, smaller, simpler days: 'Our relations with the workmen...have remained of a more intimate and personal character than is usual in concerns of this kind.' Such claims remain hard to verify, as do actual expenditures on employee welfare, although Dorman Long were probably ahead of other local employers. What *is* clear is Sir Arthur's close involvement in welfare matters, for example providing a football field for boy employees, and the detailed planning of the housing schemes at Dormanstown in 1917–18, on which great care was lavished. Equally clear is that solid embodiment of civic links, a widespread practice of employing local families over several generations. But there was a tinge of patriarchal autocracy, too. When trade union and works representatives reluctantly accepted a reduced level of war bonuses in 1922, company officials told them that Sir Arthur would have been personally disappointed if they had refused: 'He would have been very much put out, especially after his great consideration for his men and the fact that he has kept the Works running many a time simply to provide them with work.' Although labour relations on Teesside were temperate, such nanny-ish calls for simple-minded gratitude were probably already declining in credibility.[1]

Stewarts and Lloyds' welfare paternalism was less emotional, more systematic. An employee welfare fund, based on joint contributions and started during the war, distributed sickness, death and retirement benefits according to set scales. A limited type of profit sharing, modelled on that of United States Steel, started in 1925. But in the 1920s Stewarts and Lloyds showed little enthusiasm for new moves on housing, pensions or communications with employees. Its labour policies, like so much else, exhibited a characteristic financial caution.[2] On the USC's pre-1928 welfare policies still less evidence exists. It is clear that, because of the USC's wide geographical and industrial diversity, its responses to the industrial relations vicissitudes of the immediate postwar period and the 1920s represented a bigger test. Signs of moderation towards organised labour were already emerging. In 1918–20 Harry Steel was notably absent from the group of steelmasters who favoured massive deflation as 'a sharp lesson to labour'. Albert Peech was something of a Baldwinite on labour matters, as shown by his dove-ish comments on the main trade union, the ISTC; his frequent private references to labour's 'co-operativeness', 'friendliness' and 'loyalty' and to 'very difficult times for the men' in relation to wage cuts under sliding-scale agreements; and his pacific, if somewhat vague, initiatives on labour relations within the NFISM in 1926–7. After the General Strike the USC board, annoyed that steelworkers had been 'driven into wrong action', nonetheless felt that 'as in the future we should have to work with Trades Unions, our policies should be framed accordingly'. It rejected legal action against employees, apparently refrained from victimising practices like those of, for example, Consett, and showed concern at the industrial provocativeness of the government's proposed retaliatory trade union legislation in early 1927.[3]

By the late 1920s a clearer contrast emerges between the three firms. Dorman Long, stuck with its old leadership and persistently in financial trouble, was hardly in a position to take fresh initiatives on labour matters. But the new régimes in both the USC and Stewarts and Lloyds soon showed attitudes and behaviour patterns that were to be highly significant up to 1939. From 1928 onwards an interesting difference between these two firms appears. Of course, that difference was not only or even mainly manifested in labour policies where, on the contrary, much similarity existed with respect to trade union relationships, collective bargaining and wages, also with regard to calculations of what welfare measures would be useful for growth and/or efficiency. On the other hand, labour policies could sometimes test the degree of responsiveness to outside opinion and public policy, the value assigned either to wider concepts of growth and efficiency or else to *non*-growth and *non*-efficiency objectives. In so far as they did, they formed one element of a striking contrast between the USC and Stewarts and Lloyds over social action.

As was suggested in Chapter 5, the new USC régime had quite strongly developed concepts on labour matters. Robert Hilton believed, for example, that 'anybody who is "slim" in his dealings with labour is sure to be found out', and that the *first* priority for a well-educated entrant to industry was 'to study and understand the point of view of labour'. He accepted the bulk of the ISTC's proposals for reorganising the industry in 1931–2 (see below) and favoured some organised labour representation in the central policy-making process: 'We shall no longer be able to keep labour at arms' length, but will have to ask for their co-operation'...'I shall be surprised if their proposals are not far more moderate than the bulk of the steel trade believe possible.' Benton Jones's similarly large views on labour matters had a more palpably emotional and ethical content. For example, he argued that workers' pensions would help to reduce unemployment and 'add greatly to the well-being and sense of security of the working classes', as well as forming a precautionary measure against the possible legal enforcement of worker pension schemes (a question he apparently misunderstood). In his speeches, visits and contacts Benton Jones sought some kind of personal communion with employees and he aspired to social policies which clearly transcended considerations of efficiency and growth. Thus a works 'must be fit to live in', a concept which 'goes further than what is generally understood by industrial welfare'; lighting, air, heating, cleanliness and tidiness should show 'higher standards ...than has ever been reached in any works of our kind'; health conditions should be well in advance of the law; work was something in which men could and should be 'happy', clearly an objective in its own right; and the workplace was a 'community', a school for 'our relations with one another', which should develop individual human talents fully.[4]

If it was unlikely that the USC's practice would accord fully with such ideals, its policy achievements from 1928 onwards were substantial. In 1929, while still wrestling with urgent economic problems, the firm introduced a system of works councils, incorporating the principle of worker-elected representatives from every department. Hilton took a personal interest in this scheme. In 1930, for example, he expressed pleasure that worker representatives were 'keeping members of that Committee who happen to be on the Management up to their duty'. By 1931 an annual company-wide conference of works council representatives and chairmen was being held, putting the USC in the vanguard alongside leading firms like ICI. By 1934 Hilton was claiming, with some hyperbole, that the works councils 'discuss general policy of our Company, our aspirations for the future, and the difficulties which we may encounter'. In these discussions a 'hardy annual' was the employees' desire for company pensions. In fact, the USC introduced comprehensive, compulsory, joint-contributory pension schemes for staff in 1935 and, significantly, for labour in 1936. Important features were the introduction of the scheme through the joint consultation process and the inclusion of a progressive, redistributive element, with company help to ensure an attainable objective of £1 a week pension for 'the labourer who, after all, has the least opportunity of saving money during his working life'. In a letter to one of the works council chairmen Hilton claimed: 'We can feel justly proud that we are the first large undertaking in the Steel Industry to adopt a scheme of this kind.'[5]

By contrast, Stewarts and Lloyds' approach to labour matters was undistinguished: satisfactorily efficient at best, cheese-paring at worst. As was explained in Chapter 5, whereas Benton

Jones and Hilton had strong interests in this matter, Allan Macdiarmid, for all his brilliance in other areas, had virtually none.
No signs of any spontaneous interest emerge from his speeches,
private papers or practical actions, even from the recollections of
those who admired and liked him. He was content, it seems, to
leave labour matters largely to others, and his firm's record from
1926 to 1939 is bereft of major new schemes or substantial
pioneering. Although sometimes important, the changes it did
introduce were primarily reactive, imitative and lacking in
breadth.

On one front, that of trade union relationships at Corby, Stewarts and Lloyds probably adapted effectively to the requirements of cultivating the goodwill of trade union national leaders. It favoured union membership among the new employees
and developed an efficient collective bargaining framework. On
other fronts it began to lag behind, showing *lacunae* surprising
for a firm so sophisticated and relatively prosperous. Ironically,
an economy campaign in 1932–3 was needed for it to be recognised that staff recruitment methods were slipshod and outdated. There were no works councils, and pensions only for staff.
The detailed capital expenditure plans for Corby, exhaustively
discussed in 1930–3, left out general works amenities, labour
transfer costs from Scotland, a works club and church buildings
– provisions for all of which, to the extent of about £80,000, were
delayed until they became matters of extreme urgency in 1934.
Interestingly, at that late hour, the aim was described as
'amenities *such as other progressive firms have provided*' (my
italics). Despite large profit increases from 1933, company contributions to the employees' welfare fund were not raised until
1938. There is no sign of Coronation gifts to employees in 1937
(Dorman Long gave £14,000, the USC ten shillings to every
existing and retired employee, making about £15,000). Christ-
mas bonuses for staff started in 1937 'in accordance with the
merits of each individual case' (in the USC a smaller total was
distributed to all less well-paid staff). Again, a rather niggardly
scheme to refund fees and 'in exceptional cases, to pay bonuses
up to £2, on examination results for approved attendance at
educational or technical institutions', did not emerge until 1937,
considerably behind both the USC and Dorman Long. When the
Legal and General explained that staff who joined the Forces in
1939 were considered to have left the company pension scheme,
the board's reaction was not uncharacteristic. 'After a discussion', it decided 'that no commitment should be entered into
meantime, but that the matter should be dealt with if and when
occasion arises.'[6]

## Employment and redundancy policy

However, it was employment and redundancy policy that produced a bigger test for social action. The question of how far to
maintain uneconomic jobs, particularly in depressed areas, and of
how to treat redundant employees, raised wider issues of responsiveness to public opinion, question marks about social duties
towards both employees and local communities. There can be no
question that most managements found having to sack people
during the depression years a terrible trial, judging by the statements of men as diverse as Sir Arthur Dorman, Sir Hugh Bell,
Charles Mitchell, Albert Peech, Walter Benton Jones, Robert
Hilton.[7] Difficult questions arose over possible restraints in
making employment cuts, planning them carefully, showing sensitivity to public criticisms, transferring workers where appropriate and, not least, providing some compensation to those who
were unavoidably dismissed. Comparisons of company responses
must obviously take account of varying perceptions of national
interests, financial capacities, and local circumstances. That
serious conflicts were involved is shown by the sharply contrasted
experiences, once again, of Stewarts and Lloyds, and the USC.

Stewarts and Lloyds' first major test on this front was in 1931
when it bought the Scottish Tube Company with the deliberate
aim of extensive plant closures. To the master strategy for tube
industry rationalisation and massive reallocation from Clydeside
to Corby the STC formed a serious obstacle. It was a large-scale
importer of continental steel, a relatively inefficient and financially weak firm whose medium-size tube manufacturing capacity
had been running at only 50 per cent but could easily be revived.
For Stewarts and Lloyds, though, there was a difficult ethical
problem: what was the responsibility of an acquiring company for
newly-acquired but often long-serving employees whom it wished
to dismiss? The evidence suggests that the firm consistently tried
to minimise its responsibility. The acquisition was presented to
the STC shareholders and the public mainly in positive terms, as
'a further step in the rationalisation of the Tube Industry',
making the trade in Scotland 'one unit', involving 'closer co-operation' and 'presenting a united front to competitors at home and
abroad'. Only a mention of 'wasteful duplication' hinted at the
closure threat. Privately, Stewarts and Lloyds made certain
undertakings pertaining to STC's long-serving staff (not labour)
but the interpretation of these undertakings soon brought complaints from STC's chairman, H. J. Rhodes, who had become a
Stewarts and Lloyds director. In April 1932 four of the STC's
eight works were closed – Saracen, Clydesdale, Glasgow and

Union – and by May the dismissals had extended to 130 staff (over half the total) and, in all likelihood, several hundred workers. There is no evidence of extended notice or compensation to either group, although Stewarts and Lloyds' analogous closures of newly acquired plant in the Midlands in 1930, again in 1932, had at least brought some grants 'for the relief of distress', admittedly for much smaller numbers.

By October 1932 Rhodes was engaged in a rearguard action on behalf of the particularly long-serving staff (again, not labour) among those who still remained. He contended that they would have obtained a year's salary if they had been immediately sacked in January 1932. This they should still receive since 'it had suited Stewarts and Lloyds' convenience to retain their services up till now'. In a letter to Macdiarmid Rhodes accepted this was a question of the spirit rather than the letter of the original undertakings, pleading for 'some modified form' of a redundancy compensation scheme currently being discussed for Stewarts and Lloyds' staff. But there is no sign of a reply from Macdiarmid and the only concession was an extra one or two months salary to the very long-serving. Letters from some of the latter continued to allege markedly worse treatment than that currently meted out elsewhere in the firm. A final smaller episode again reflected Stewarts and Lloyds' minimising approach. In a slight gesture towards social responsibility, the STC's welfare supervisor had been 'loaned' on his full salary of £280 to the Coatbridge Unemployment Social Centre. In April 1933 Stewarts and Lloyds' central committee tried to arrange a company subscription of between £100 and £150 to the Centre 'subject to their taking the STC Welfare Supervisor into their own employment'. By May, when the man was finally about to leave the firm, the committee agreed to make 'a subscription not exceeding £100...at a later date *if we are asked to subscribe*' (my italics). The attitude of waiting to be pushed was not untypical.[8]

A further and larger phase occurred with the run-down of Stewarts and Lloyds' own long-established interests on Clydeside. The announcement of the Corby scheme in late 1932 led to widespread local fears about a possible abandonment of tube making in Scotland, a subject on which, according to the *Glasgow Herald*, 'the directors of Stewarts and Lloyds are silent in their circulars'. The firm immediately denied any intention of 'abandoning tube manufacture in Scotland', which was literally true, if anodyne, adding misleading assurances, however, about 'transfers of work from Scotland to England and *vice versa*' and hopes for '*more*, not less work for our Scotch plant' (my italics). In May 1933 Macdiarmid's annual speech, which hinted more openly at reductions while repeating that 'there is no question of our withdrawing from Scotland', was followed the next day by an announcement of the closure of Stewarts and Lloyds' Vulcan Works at Motherwell which normally employed 200 people. But this was only a prelude to a larger operation involving the historic Clydesdale iron and steel works. Here a dovetailing of Stewarts and Lloyds' Corby plans with a wider scheme for reorganising the Clydeside iron and steel industry involved a long wrangle with Colvilles and its chairman, John Craig. Under the terms of the final agreement, announced in October 1933, Colvilles were to take over Stewarts and Lloyds' plate business in Scotland while Clydesdale's existing plate mill was to be reconstituted as a blooming mill and transferred to Corby. In late 1933 the consequential closure of the Clydesdale works at Mossend, involving some 800 workers, led to widespread local protests.

This time Stewarts and Lloyds' social consciousness was greater. Substantial numbers of the workers made redundant were encouraged to move south in order to take up new jobs at Corby and some £21,000 had been set aside for allowances for long-serving dismissed staff. However, it is not clear how much of this was for statutory obligations or indeed how much of it was actually spent. Some ambiguity also remains on three further points: (1) whether any compensation was paid to long-serving redundant workers who were unable or unwilling to transfer to Corby, especially, perhaps, those who remained out of work; (2) whether any help was given towards wider local efforts to improve Clydeside employment and relieve pockets of extreme distress; and (3) why the burden of public explanation seems to have fallen largely on the shoulders of John Craig and Colvilles. With regard to (3), public concern extended from Clydeside even to the BID Co, normally remote from social issues, where Charles Bruce Gardner wrote about the 'poor unfortunates at Clydesdale' in an effort to reassure Sir Andrew Duncan. But from the private as well as the public discussion Macdiarmid and Stewarts and Lloyds were, it seems, markedly absent. It should be added that by late 1933 Stewarts and Lloyds' profit trend was already strongly upwards, thus leaving more room for manoeuvre in dealing with social problems. On the available evidence, the manner of the firm's retreat from its historic involvements with Clydeside hardly seems distinguished.[9]

The problems are thrown into even sharper relief by the USC's markedly different experience. As already mentioned, that firm was committed, by contrast, to an overall strategy of conservation and economic salvaging, a vital element of which was relatively stable employment. Large-scale lay-offs and redundancies occurred elsewhere in the industry, particularly by 1930–2, leading to vast improvements later. Dorman Long, for example, increased its employment by over 50 per cent between 1932 and 1936. But the USC's employment size showed little reduction through the worst phases of the depression: early 1920s, around 25,000; 1931, 26,000; 1939, 31,000.[10] To a major extent this reflected the policy of maintaining the firm's biggest trouble-spot, its West Cumberland activities, where drastic reductions could have lopped off thousands (for the chequered history of the USC's earlier involvement in the area see Chapters 3 and 4). It is necessary to explain the substantial debt which the continued sustenance of West Cumberland owed to social action.

During the 1920s the local press frequently referred to the area's acute dependence on the USC and to fears of large-scale redundancies or even a USC withdrawal. The local authorities pleaded with the firm to maintain jobs and to exert political pressure to get overseas rail orders for the local works. The firm did both these things while accepting, both privately and in public statements, that its protective role was crucial. When the new top management took over in early 1928 local fears reached a new climax. Would a new, tough-minded régime, facing a grave crisis, empowered to wield a new broom and urgently seeking a complete financial overhaul, finally cut out the painfully weak West Cumberland units? As we have already seen (Chapter 7), it did not; but the reasons for continuing, indeed sustaining these activities were complex. Although important, economic expectations were extremely modest. At best only a small profit from West Cumberland was envisaged and one largely dependant on uncertain extraneous factors. If this was a necessary condition for the policy, it was hardly sufficient in view of the USC's acute crisis. In fact, there was already a built-in assumption that there was no alternative, morally and politically, but to stick things out and throw in extra resources. Although the USC board probably included some sceptics, there is no sign of that assumption being seriously questioned. It owed much to sensitivity on the issue of local employment and this, in turn, seems to have derived from several factors: Benton Jones's concept of social responsibility, and sympathies stemming from his experience of colliery areas; Hilton's similar distaste for the role of a ruthless butcher; managerial awareness of local impacts and links with local labour resulting from frequent visits; a sheer dislike of public opprobrium. Even the view that West Cumberland could be salvaged commercially may have owed something to these social attitudes and in that sense could have been a partial rationalisation. Within three months of the early, critical decision to sustain West

Cumberland in summer 1928, Benton Jones publicly acknow-
ledged the social obligation, explaining that the Workington
Branch 'finds direct employment, even in these times, for 7,700
men, and indirectly the whole countryside is dependent upon
it'.[11]

What followed was a large, sustained exercise of discrimination
on Workington's behalf. Although some economic improvements
emerged during the 1930s, profits were still elusive. Even in 1939
a consultant's report reiterated the toughness of the area's economic problems for the USC and the long-term nature of further
efforts to make it viable. Yet large sums were spent on investment projects. During the years 1928–33, when the USC's
finances were still strained and West Cumberland was still
making outright losses, its share of total group capital expenditure was 34 per cent: over the whole period it took a dispropor-
tionate share relative to average profits. Enormous efforts were
thrown into diversification projects, with varying success. Management pressured other branches to prefer Workington supplies
over other sources, allowed Workington to undercut them in
third-party markets and allocated orders in its favour, for example for billets. Much of this exacted a price in terms of resentment
from other branches. Resources of management time, whose
opportunity costs were high, were disproportionately devoted to
West Cumberland problems. As the steel industry passed into the
new régime of national price regulation and collective buying and
levies in the late 1930s, the USC had to undertake advocacy of
the area's interests at the national level. There were intra-USC
problems on price policy, notably when the combine favoured
stabilisation but exceptions had to be sought for Workington
products. Overall, the diversionary effect from other possible
priorities, for example for strategic growth, can hardly be doub-
ted.

If anything, the role of social considerations in this policy
probably increased during the 1930s. As public opinion and government became more concerned about unemployment in the
Distressed Areas, it became clearer than ever that a USC withdrawal would blatantly defy the social consensus, and the firm
continued to show sensitivity to such pressures. Visits from
Sheffield, public reassurances and the local press were all used in
efforts to allay local fears. The Workington general manager
through the 1930s, Robert Crichton, played a leading part in civic
efforts to relieve unemployment and to champion the area nationally. Certain secondary aspects of the West Cumberland policy
reflected elements of public duty, notably a decision in 1939 to
spend £139,000 on an iron ore mine which, although unlikely to

be remunerative, was needed for national strategic purposes. Slightly later, Benton Jones argued for maximum regularity of employment as the key policy for Workington, in private. Equally in private 1938 correspondence from the American consultants, Arthur McKee confirms that withdrawal was regarded as 'quite out of the question' because of both the capital invested and 'the number of workmen dependent on it's [Workington's] operation'.[12]

### Philanthropy and pollution

Before turning to wider public policies it is necessary to refer briefly to two other issues relating largely to local communities. The first, already hinted at, is philanthropy, which went over-whelmingly to local activities. Although relatively minuscule, usually something under one per cent of profits, philanthropic flows could have important local effects and are certainly managerially revealing. Typical features include boardroom decision-making about individual gifts and a close involvement of the firm's leading figures. Thus we find J. G. Stewart deciding on a subscription to a war memorial fund in 1921, Sir Arthur Dorman fussing about gifts to this or that educational body in Middlesbrough, Sir Hugh Bell writing in 1927 about churches in Kent ('I should rather like to reply to the Archbishop before I leave England'). Varying managerial proclivities emerge, for instance Stewarts and Lloyds' preference for system and regularity, Dorman Long's special concern for ex-officers, war disabled causes and war memorials (Sir Arthur's eldest son had died in the Boer War), and for small emergency gifts to foreign lands like Austria, Japan and India. During the 1914–20 period overall philanthropy tended to rise only to diminish in the 1920s, as one would expect. Dorman Long, however, remained relatively open-handed. Through the 1920s it regularly subscribed several hundred pounds per annum to bodies like the Cleveland Technical Institute, Constantine Technical College, Middlesbrough YMCA, North Ormesby Hospital and North Riding Infirmary, also assisting, for example, Lord Haig's Appeal (£600), a local parish hall (a 3 per cent £400 loan), Durham Voluntary Hospitals (£500), the National Industrial Alliance (£200) and Dormanstown church (£1,500 guarantee).[13]

But again the new managerial régimes brought an increasing polarisation. From 1931 onwards Dorman Long's philanthropy appears to diminish. Once more the main contrast is between the USC's breadth and Stewarts and Lloyds' minimising. The USC made a series of sizeable contributions between 1934 and 1938 as financial conditions eased, reflecting both its geographical spread and a wide range of interests: £1,500 each for Frodingham and Thurcroft churches, £2,500 for Scunthorpe War Memorial Hospital, £1,000 for the National Thankoffering Fund, £2,000 for metallurgy at Sheffield University, £2,500 for Workington social services, £2,000 for Workington Infirmary, £500 to the Bishop of Sheffield's retiring fund, £500 for the Sheffield YMCA, £750 for Rotherham Hospital, £1,600 for Sheffield Voluntary Hospitals, £500 for Old Brumby Methodist church, £250 for a fracture clinic in Whitehaven. By the late 1930s the USC's philanthropy, running at several £thousands each year, represented the best part of one per cent of average profits and it was highly profit-elastic, roughly trebling between 1936 and 1939 while profits increased only moderately. By contrast, Stewarts and Lloyds' individual charitable grants tend to drop out of the Directors' Minutes from 1926, the customary gifts of newly-acquired subsidiaries were reduced as the latter were closed down, and there are no signs of major gifts to parallel the USC's. By the late 1930s Stewarts and Lloyds' carefully tabulated charitable contributions were no more than about £1,000 per annum, a small fraction of one per cent of the firm's profits and lagging far behind the latter's rapid growth. Here, then, was a marked contrast between approaches to social action, quantitatively tiny but, perhaps, qualitatively significant.[14]

The second largely local issue was environmental pollution. Here, too, iron and steel firms were sometimes faced with the question of putting a value, in effect, on both social goods and their own public repute. The main problems were that public opinion's signals on environmental issues were still weak, that the causes and effects of pollution were often genuinely controversial, and that methods of alleviation or prevention could correspondingly be uncertain as well as potentially highly costly during years of recession, and also hard to organise on a voluntary collective basis where firms of varying sizes, finances and social attitudes were involved.

Most of these difficulties applied to the issue of chronic pollution of the River Tees. Successive investigations from 1923 by the Ministry of Agriculture and Fisheries led to official criticism of the main heavy industry firms on Teesside, including Dorman Long. Suggestions followed for a modest, jointly-financed research programme, 'a cheap insurance', it was said, against harassing correspondence about real or alleged damage to fish. Dorman Long joined with the other firms in obstructing and criticising this scheme while providing a trickle of funds for it. In

1928, in common with the others, the firm was still resisting
further industry financing of the scheme: 'It was not felt that the
local Ironmasters should pay to prepare a rod [i.e. legislative
penalties] for their own backs.' To be fair, the long-accumulated
pollution of the river was a specialist concern which occasioned no
great public outcry.

Some of the nuisances were richly disputatious, for example the
effects of effluent slag on the Redcar marshes on the river three
miles away and the nicely arcane question of whether drains
discharging through sluices from marshy lands were 'tributaries'.
Moreover, Dorman Long and the others were able to argue that
much of the pollution was from the town, not industry, and so
ought to be a shared responsibility with the local authorities. The
issue dragged on through the 1930s with, it seems, only slight
improvements.[15]

By comparison the externality issues surrounding Stewarts
and Lloyds' massive incursion into Northamptonshire, although
more diverse and widely controversial, were at least relatively
simple. The creation of a large industrial complex in a deeply
rural area was bound to produce highly mixed spillovers. Some of
these were obvious social 'goods' (increased local income, help
with local unemployment), some were 'bads' (alienation of farming land, factory effluent). Others were ambivalent and liable to
conflicting local evaluations, notably the incursion of large numbers of Scots, the appearance of the new town at Corby, the
accelerated drift of local labour from farming, the transformation
of a nearby valley into a lake owing to vastly increased water
requirements. Much of this, anyway, Stewarts and Lloyds could
do little to alter, whether for good or ill. Where the situation *was*
malleable, its actions were sometimes purely defensive: for example, a local rating assessment of £40,000 was characteristically
contested through the courts. On other matters the firm adapted
quickly to local criticism or glaring social needs (a careful culti-
vation of the local press, measures to stop pollution of local
streams) or made tangential pacifying moves (a supply of softened water for local housewives, talk of steel tubes protecting
trees from gales). Certainly, its employees spontaneously began
to make distinctive contributions to local sport, recreational and
cultural activities.[16]

On one striking matter, however, Stewarts and Lloyds' impact
in Northamptonshire partially justified those who regarded it as
an 'industrial cuckoo in a rural nest', a perpetrator of 'gaping
wounds'. This was the issue of land spoilt by open iron ore working. Pre-1914 methods of hand-working had caused little damage
and for many years the restoration of land for agriculture had

been relatively easy. With the introduction of large digging, dumping and excavating equipment, the problem worsened after the First World War. An official commission, appointed by the Minister of Health and chaired by Lord Kennet, reported in 1939 that 'the position became more serious' when Stewarts and Lloyds, as large-scale newcomers, had to lease areas of deep overburden. This had caused much spoliation which the firm did little to remedy. It was at Corby, in fact, that 'the greatest devastation has occurred'. Mainly in connection with Stewarts and Lloyds, around 3,000 acres had been spoilt for agriculture by 1937, of which only 500 acres had been afforested, but not levelled, and 250 acres was used for dumping of waste slag, etc. The report commented on 'wide expanses of devastation', 'the miserable impression of ruin and waste', 'the multiplication of rats and other vermin', proliferating weeds, and the cumulative threat to Northants at projected rates of working of the 60 per cent of its total acreage that was ore-bearing. Unfortunately, the quantification of remedies in the report was weak, partly due to the unwillingness of the firms to provide full data on costs. Little confidence was expressed in voluntary action on levelling and afforestation, and the recommendation was firmly for statutory changes. Overall, it was a disturbing commentary on some neglected social costs, particularly of Stewarts and Lloyds' greatest venture of the 1930s. Even the firm's official apologist, writing some thirty years later, could find little effective to say in mitigation.[17] Perhaps the best case to be made for the firm is that it was in a tremendous hurry to get economic results from Corby, that the nature of the best remedial measures was not unambiguous and that public opinion was not yet fully alerted on the environmental issues; but it *could* have behaved differently.

## Notes

1  D/AGM, 17.4.23. D/DM, 7.1.19, 13.7.20. Papers on formation of Dormanstown Estate (1066/21/4). Interviews with retired D L officials. Britannia Works (803/1/2), meeting, 19.5.22. For references to employee pensions see D/DM, 7.9.20, 13.9.21, 11.10.22, 15.3.27, 19.6.28, 22.9.31. For miners' welfare see D/CC.

2  S/BM, particularly 20.5.14, 14.6.23, 27.3.24. S/GPC, 12.3.19, 3.3.19, 3.5.23. S/AGM, 28.3.25. S/EC, Data on Employees' Benefit Fund, 1922. S/CB for annual contributions to latter.

3  F/GM, 17.7.19. U/DM, 8.7.24, 12.5.25, 8.6.26, 11.1.27, 8.2.27, 9.8.27. U/AGM, 11.10.21. NFISM, Parliamentary and General Purposes Committee (01547), 16.2.27. For Consett's post-1926 militancy see S. Tolliday, 'Industry, Finance and the State' (Cambridge, Ph.D. 1979).

4  U/MD, R. S. Hilton to Sir W. Larke, 8.6.31, to Sir Arthur Duckham, 26.1.32, to J. A. Gregorson, 19.3.34 and 29.3.34, to W. J. Brooke, 17.4.34, to

## 180   Business Policies in the Making

A. Boyden, 3.10.35, to Miss D. M. Hughes, 1.1.37. U/CC, 8.1.36. Interviews with retired USC officials. Sir W. Benton Jones, 'Three Elementary Things' (094/1/1).

5   U/MD, 12.9.29, 5.7.30, 6.12.30, 19.3.34, 5.10.34, 29.10.34, 24.8.35, 2.9.35, 10.9.35, 18.9.35, 26.9.35, 3.10.35. U/DM, 26.2.36, 15.9.36. U/CC, 8.1.36, 9.9.36.

6   S and L, Iron and Steel Trades Employers' Association (127/1/1), correspondence and minutes of Corby Labour Committee, 1934–8; also Lancashire Steel Corporation (002/2/4), notes, 1.12.38. Economy Commission, Report, 22.9.32 (1791/1/16), and Report to managing committee on staff recruitment and training, 11.5.33, and previous notes (1791/1/21). S/BM, 26.3.34, 24.4.34, 29.5.34, 2.4.35, 29.6.37, 27.7.37, 15.9.38, 6.12.38. BE: SMT 3/230, Corby progress reports, 1st Report, 30.5.33. S/GPC, 28.11.29. S/MC, 20.6.33, 31.7.33. S/CB. S/EC, November 1937. D/DM, 8.12.37, D/FC, 12.3.31. U/DM, 26.5.37, 24.12.37.

7   D/AGMs, 15.12.21, 18.12.30, 16.12.31 and *The Times*, 27.9.27; U/AGMs, 11.10.21, 23.10.28; interviews with retired DL and USC officials.

8   S/BM, 10.11.31. S/MC, 16.2.32, 26.2.32, 8.3.32, 26.4.32, 8.6.32, 4.10.32, 22.11.32, 28.12.32, 24.1.33, 21.2.33, 25.4.33, 16.5.33. Scottish Tube Company Managing Committee, in S/MC, vol.2, 16.3.32, 19.4.32, 11.5.32. SL Economy Commission (1791/1/21). Notes and correspondence on STC staff, 1932–33. *Glasgow Herald*, 17.12.31.

9   *Glasgow Herald*, 30.11.32, 1.12.32, 3.5.33, 4.5.33, 21.10.33, 23.10.33, 24.10.33, 3.11.33, 4.11.33, 6.11.33, 8.11.33, 23.11.33, 4.12.33, 7.12.33, 8.12.33, 15.12.33, 28.12.33, 30.12.33, 7.4.34, 14.8.34, 10.11.34, 15.11.34, 20.2.36, 17.12.26, 4.3.36. Colvilles, Mr Craig-Stewarts and Lloyds (2653), progress report 15.3.33, J. Mitchell to A. McCance, 17.3.33, J. Craig to A. C. Macdiarmid, 22.6.33, J. Craig to W. Stewart, 26.7.33 and 19.8.33, Craig to Macdiarmid, 15.9.33. BE: BID 1, 85/1, C. Bruce Gardner to Sir A. Duncan, 27.10.33. S/CB. There is no mention of the Clydeside retreat in S/BM or S/MC. Nor in F. Scopes, *The Development of Corby Works* (1968).

10   For total employment data see D/AGMs, 11.12.35, 9.12.36, 8.12.37, and Managing Director's Monthly Reports, 1936–38 (1066/16/11). USC General Matter file, 1924–9 (006/10/3), memorandum to A. O. Peech, 14.2.24; U/CC, 2.6.31; U/DM, 14.5.34; U/CI, 4.8.37 and 24.6.39.

11   *Carlisle Journal*, 4.1.19. *West Cumberland News*, 27.6.21, 1.10.21, 19.1.24, 10.10.25, 30.10.26, 31.8.29, 15.2.30, 5.4.30, 12.11.32, 1.6.35. Workington Borough Council Minutes, 27.9.23. U/AGM, 23.10.28. For fuller details see J. S. Boswell, 'Hope, inefficiency or public duty? The USC and West Cumberland, 1918–39', *Business History*, Jan. 1980.

12   U/DM and U/CC, 1928–39 particularly 17.9.28, 12.9.30, 9.12.36, 13.1.37, 10.2.37, 10.3.37, 13.10.37, 17.9.28, 10.11.37, 9.11.38, 10.5.39. Andrews and Brunner, interview notes, 1950. Report by Arthur McKee and Company, 1939, kindly lent to the author by Mr J. Lancaster. Andrews and Brunner, op. cit., pp. 208–9, 225, 244, 274. *West Cumberland News*, 31.8.29, 15.2.30, 5.4.30, 12.11.32, 20.5.33, 1.6.35. J. Jewkes and A. Winterbottom, *An Industrial Survey of West Cumberland and Furness* (1933), pp. 4, 79, 80, 89. J. S. Boswell, op. cit.

13   D/DM, U/DM and S/BM from 1914. S/BM, 1.12.21. S/GPC, 9.3.21. D/CI (1066/5/5), Sir H. Bell to Byrne, 23.12.27. D/DM, 17.6.19, 12.10.19, 9.3.20, 11.1.21, 13.11.23, 12.2.24, 9.4.24, 14.12.26, 18.12.28, 22.1.29, 23.4.29, 18.2.30, 18.11.30.

14   D/DM. U/DM, 14.9.34, 24.10.34, 24.4.35, 22.1.36, 15.9.36, 25.11.36, 28.4.37, 23.6.37, 22.9.37, 22.9.37, 24.11.37, 23.3.38, 27.4.38, 23.11.38, 26.4.39, 5.7.39, Summary of suggested annual subscriptions and donations, 22.11.39. U/CC,

20.9.29. S/BM. S/MC, 10.1.33, 16.5.33, 31.7.33, 14.11.33, 19.12.33. S and L, Subscriptions and Donations, Approved Lists, 1935–9 (002/2/25). S/CB. The contrast may also have applied inside S and L. Compare obituary references to benefactions as between J. G. Stewart, *Glasgow Herald*, 2.3.25, and A. C. Macdiarmid, *Glasgow Herald* and *The Times*, 15.8.45.

15 DL, Pollution of River Tees (212/1/2). River Pollution (210(b)3). Chief Chemist's letter books, 26.7.24, 19.9.24, 24.9.24, 7.11.24 (203/1/77).

16 *Kettering Leader and Guardian*, 10.1.36, 24.1.36, 31.1.36, 7.2.36, 6.3.36, 14.3.36, 3.4.36, 24.4.36, 15.5.36, 22.5.36, 5.6.36, 19.6.36, 26.6.36, 3.7.36, 10.7.36, 7.8.36, 11.9.36, 18.9.36, 25.9.36.

17 This paragraph is based largely on study of numerous references in the *Kettering Leader and Guardian*, 1936–8, and the *Northamptonshire Evening Telegraph*, 1938. For the local rates contest see latter, 21.10.38 and 18.11.38. J. C. Carr and W. Taplin, *History of the British Steel Industry* (1962), pp. 583–4. Scopes, op. cit., pp. 114–20, 128–31.

18 Report of the commission on the restoration of land affected by iron workings, HMSO, 1939. *The Times*, letters, 14.12.32, 19.12.32, 23.12.32, 28.12.32. 'Industry comes to Corby', *The Listener*, 13.3.35. Carr and Taplin, *ibid.* Import Duties Advisory Committee, Report on iron and steel industry, June 1937 (Cmd. 5507). Scopes, op. cit, pp. 123–4.

# 9 Relationships with public policies and government

Between the First and Second World Wars the iron and steel industry went through a socio-political transmutation as well as an economic one. Paralleling the shift from depression to economic resurgence there was a movement more subtle, and in some ways more significant, for the development of a mixed economy. This was the way in which the industry passed from privatism to politicisation, from individualism to a form of mixed, public –private collectivism. So far did this tendency progress that by the late 1930s British iron and steel exemplified a largely decentralised and informal control of private enterprise. It became the seed bed of a new and far-reaching experiment in the political economy of the 'Middle Way'.

Of course, the inadequacy and ambiguity of these terms illustrates the lack of an exact description or clear theoretical understanding of what happened: a continuing problem in this area of business behaviour. Only by carefully breaking down the phenomenon into various elements can the groundwork be laid for the full evaluation that is still badly needed. This chapter can only provide a limited contribution.

In fact, the industry's socio-political change had several dimensions. On one level it moved much faster towards greater concentration and, particularly after 1932, cartelisation. On another level the industry shifted towards a large measure of self-governing, industry-wide collectivism, focusing on a representative central body. On their own these processes could simply have meant a mass accumulation of sectional power, a jump from atomistic free enterprise to oligopoly, cartels and private federalism. But more than that was involved. For one thing, these processes themselves owed much to public policy and government influence. It is a fallacy to suppose that the industry as a whole was eager to embrace cartels and a strong central body. On the contrary, considerable public efforts were required to secure these objectives and then to maintain them. Secondly, other economic interest blocs surrounding the steel-masters were also advancing in power, notably the coal industry, the motor industry, the trade unions, the government purchasing agencies. Third, the industry became subject to more surveillance from public opinion through Parliament, the press, various official or semi-official enquiries, and opinion-formers generally. Fourth, specific government influences increased mainly through two intermediate agencies: a new body, the Import Duties Advisory Committee or IDAC (from 1932); and the industry's reconstructed national federation (from 1935). Thus the passage towards private collectivism was enveloped to a large degree within a surrounding framework of both informal socialisation and increased public power.

Of course, politicisation did not develop continuously, let alone smoothly. In the 1920s there was virtual government inaction, a continued political deadlock over the tariff issue (see Chapter 2). Also the Depression and a general conservatism helped to discourage even limited progress towards cartels and a strong central body. Between 1929 and 1931 things livened up as attention focused on the idea of giant amalgamations assisted by the BID Co. Some useful action resulted of a strictly limited kind. Then the advent of a National Government in autumn 1931 at last foreshadowed some kind of wider solution: a compromise whereby the grant of tariffs would be linked with attempts at collective reorganisation inside the industry. Between 1932 and 1934 leading figures in the NFISM struggled to produce a reorganisation scheme that a majority in the industry could be induced to accept and which would at the same time be acceptable to IDAC – hence also the government – as a *quid pro quo* for continued high-level tariff protection. In this process ambitious ideas for plant rationalisation, area mergers, trade association reform and a central body with large powers all fell by the wayside as major concessions were made to the industry's conservatives.

What emerged by 1934 looked like a second- or even third-best solution: a plan for a mildly strengthened central federation equipped to bargain with foreign interests, encourage cartelisation and provide common services against the background of some continuing supervision by IDAC. But this apparently emaciated and anti-climactic régime soon developed greater vigour than might have been expected. Under the protective-*cum*-critical aegis of IDAC and led by an arch public–private intermediary, Sir Andrew Duncan, the new British Iron and Steel Federation (BISF) pursued far-reaching experiments in voluntary price control, investment supervision, collective buying and selling, and central policy-making, within a continued structure of private enterprise.

Both the long-term movement towards this middle way and the degree of its success from 1935 onwards demanded social action. The individual firm needed to accept (1) a collective concept of growth and efficiency, and (2) certain wider public duties. Item (1) ruled out an adversarial, zero-sum approach and suggested that the growth–efficiency–profit interests of all would be better served, at least in the long-run, by co-operation. Arguably, industry-wide co-operation would secure external economies of scale in research, international cartel bargaining, export selling, bulk buying of imported materials. It would also help to avoid wasteful duplication in investment as well as evening out investment–price–profit fluctuations. But this collective enhancement of sectional interests implied duties to those *outside* the industry. Under (2), then, came ideas of social obligation, prudential exchange and public repute: that a centralised cartel should not exploit its power, that government support for it, together with continued tariffs, called for accompanying duties without which such benefits might be withdrawn, and that the industry's reputation and freedom from political interference also depended on its behaving moderately. Both (1) and (2) demanded some surrender of powers to the central body. They could suggest consortia and joint projects where a firm might prefer to go it alone. They often required immediate sacrifices – levy contributions; loss of opportunities from unilateral action, for example on prices by a trade association or on investment by an individual firm; higher prices in a recession than some would prefer, lower prices in a boom.

In theory, firms could react in three principal ways: co-operation, passivity or separatism. A co-operative firm would accept the collective concept of growth and efficiency, and its corollary, the central body's interpretation of 'the good of the industry'; also the wider idea of public duties and hence the central body's need to compromise with some external public supervision. In neither case would mere conformism be desirable or feasible. On the contrary, co-operation would imply an active contribution to collective policy-making, possibly including some vigorous disagreements – in fact, a participative mode. By contrast, a passive firm, primarily influenced by custom, habit and majority opinion, would broadly conform with the prevailing régime, whether of privatism and limited cartels up to the early 1930s or the later increased cartelisation, centralisation and politicisation. Finally, a separatist response would mean a continued, active determination to put individual corporate interests foremost. If industry or public interests could be harnessed to these, well and good. If not, the industry or public interests would be defied or evaded.

Obviously, these are polarisations and for a firm or even an individual to fit completely within one of the three categories would be rare. However, it will be argued in this chapter that co-operation was the principal response of the USC, passivity of Dorman Long, separatism of Stewarts and Lloyds. Again, this does not necessarily mean that each firm's stance was consistent or that it represented a polar extreme within its own category. For example, in the early 1920s the USC leadership was not as advanced in industry-wide thinking as, say, John Craig of Colvilles; Dorman Long were probably never quite so passive as many relatively obscure firms; and Stewarts and Lloyds' separatism was probably less extreme, certainly more sophisticated, than that of some other dissident firms in the 1930s.

## Dorman Long's passivity

Let us start with Dorman Long's basic characteristic of public–political passivity. At no point during the interwar period did this great firm either adamantly obstruct or zealously assist the tendencies just discussed. While others made the running in either direction or even occasionally, by turns, in both, Dorman Long moved fairly consistently along a cautious middle road, shifting with the main tides. Thus at the start it was an orthodox, limited carteliser, conservative on industry-wide reforms, and at the end an orthodox, limited conformer within the new highly organised régime. That this basic continuity went alongside the dramatic economic and managerial changes chronicled in earlier chapters is at first sight paradoxical. But it is important to understand particularly how, against the background of the firm's persistent economic problems, the distinctive characteristics of four men – Sir Arthur Dorman, Sir Hugh Bell, Charles Mitchell and Ellis Hunter – contributed in very different ways to broadly the same result.

Sir Arthur Dorman had long publicly advocated fairer treatment of the industry by government (in the shape of tariffs) and the capital market (in terms of finance for investment). As an ardent imperialist and a Conservative tariff reformer, he was no stranger to public discussion. However, on the postwar issues where steelmasters themselves could do something collectively to influence the situation his response was far from vigorous, even though Dorman Long was constrained by other firms on the north-east coast, including South Durham and Consett, which were, if anything, even more suspicious of collective action.

Thus when a departmental committee considered the postwar

position of the iron and steel trades between 1916 and 1918,
Dorman Long failed to advocate any radical reorganisation regio-
nally or nationally. It went along with the north-east coast
makers' extremely weak evidence to the committee, which
bewailed the collective, large-scale organisation of foreign competitors but illogically advanced no constructive counter-measures.
When John Craig argued within the NFISM executive committee
in 1922 for 'a General Staff for the Industry', common export
selling, co-operative purchasing and 'a united front' against foreign competition, Sir Arthur did not join the USC's Albert Peech
in supporting him. Dorman's presidency of the NFISM in 1923–4
was marked by no new initiatives on central organisation. In the
north-east he dragged his feet on a scheme for a common export
selling company, although after much price warfare and wrang-
ling the region's leading firms did improve their arrangements by
the late 1920s. His annual chairman's speeches showed no enthusiasm for either cartel-tightening or rationalising mergers (see
Chapter 4). His ideas of general economic revival, tariffs without
political strings and moderate (unsupervised) cartelisation were
probably representative of mainstream opinion within the industry.[1]

To all this Sir Hugh Bell's greater political prominence and
rather different ideology made little difference. Indeed, Sir
Hugh's idiosyncrasies contributed at least indirectly to Dorman
Long's basic passivity during the 1920s. His fervent adherence to
free trade, a signal triumph of ideology over corporate economic
interests, blurred the firm's public image on the tariff issue. His
intellectually consistent and equally passionate espousal of Manchester School liberalism on other issues may have contributed to
Dorman Long's ambiguity on competition and cartels. A rigid
adherence to 'absolutely inevitable economic laws' and a blanket
excoriation of government interventions of any kind ensured a
high political profile which was often tangential to the main
problems facing iron and steel. There was so much of the elderly
Don Quixote about all this that it hindered Dorman Long from
either supporting the new currents or even very successfully
obstructing them. If Sir Hugh had survived 1931 he would certainly have found it hard to stomach the mixed economy experiments of the 1930s.[2]

Of course, Charles Mitchell was unconstrained by the conservative doctrines of his two predecessors. He was probably an
economic pragmatist, certainly an adherent of rationalisation via
mergers. As a newly arrived leader in the industry, an ambitious
man who partly owed his position to the BID Co-led reorganisation current of 1929–31 (see Chapter 5), Mitchell was likely to

play in with national-level moves towards reform. What happened, though, was both more dramatic and ambivalent. Mitchell was drawn personally into central reorganising efforts to such an extent and with such confused results that once again the net result was to confirm Dorman Long's passivity. In spring 1932 the newly formed IDAC, set up by the government to monitor the establishment and variation of tariffs and related issues of rationalisation, decided that the steel industry's *quid pro quo* for high tariffs, a decisive move towards collective reorganisation, would best be initiated by a national committee formed of (carefully chosen) leading representatives of the industry. In searching for a chairman they lighted on Mitchell, who had impressed the City and who was utterly committed to a show-piece rationalisation project, the Dorman Long–South Durham merger scheme.

Mitchell's chairmanship of this controversial National Committee from 1932 to 1934 probably tied Dorman Long's hands politically. On the one hand, it left the firm unable to pursue a distinctive line on the reorganisation issues where it might well have been, in the main, more conservative. Thus the firm kept a low profile in the great NFISM debates about reorganisation, unlike its north-eastern colleagues, South Durham and Consett. Quietly, it seems to have followed the pack towards the final compromise of 1934. On the other hand, in his efforts to find a reorganisation scheme acceptable to both the industry majority and the IDAC Mitchell, alongside his committee colleagues, was forced into one tactical shift after another. His committee first showed signs in 1932 of merely reflecting the existing balance of industry opinion, basically for tariffs plus cartels but no strong central organisation and no public responsibility strings. Next, pushed hard by the IDAC, the committee swung to the other extreme in early 1933 with a scheme for a new central body with a full-time independent chairman, which would deploy a central fund, promote trade association reform and 'assist' on mergers and redundant plant. But when this scheme aroused major dissension within the NFISM Mitchell and Sir Alfred Hurst of IDAC watered it down in order to get majority acceptance which finally emerged by spring 1934. The whole process left the unfortunate Mitchell looking somewhat pushed about and did not enhance a public reputation already battered by Dorman Long's difficulties and the South Durham merger débâcle.[3]

By 1935–6 when Ellis Hunter was coming to the fore, the new BISF, under Sir Andrew Duncan's independent chairmanship, was also well established, so there was less chance to influence the nature of the new régime. The main question was how far

Dorman Long would fall in with it. Some of the firm's reactions were still distinctively conservative, as when in late 1935 Laurence Ennis privately agreed with Sir Charles Wright's objection to a unified costing scheme for the industry: 'After all, it is the business of the individual members.' In March 1937 Dorman Long refused to supply the BISF and IDAC with information regarding sheet-bar production costs for price control purposes while the figures it did provide were described by IDAC as inflated. Under pressure from Sir Andrew Duncan, the firm gave way.

But these incidents were minor compared with the deviations of, say, Whiteheads or Stewarts and Lloyds (see below). Once again, Dorman Long seems to have conformed with mainstream big firm practice, by now largely playing along with collective rules. For example, in September 1935 the board felt that any new scheme for a cold rolling mill for the north-east should be not only 'in close consultation with other North East Coast manufacturers' (a familiar point of attempted liaison particularly, even at this time, with South Durham), but also 'in collaboration with the Federation as regards the national production of this country'. Only the BISF's 1937 scheme for central purchasing of foreign scrap and pig iron, financed by a universal levy but with rebates to firms receiving little benefit, led to some argument. Dorman Long felt it would be disadvantaged as a firm which mainly relied on its own supplies. Otherwise no hint of dispute emerges from the BISF, IDAC or Dorman Long records. But then the new Ellis Hunter régime had its work cut out on the home front and needed to restore Dorman Long's battered reputation. It had no strong economic reason for taking issue with Duncan or IDAC and probably particularly welcomed the partial taming of its main local competitor, South Durham. In any case Hunter, a man destined to grow industrially and politically, was far from being a pure sectionalist or a natural dissident.[4]

## The USC's co-operativeness

If a mixture of economic tribulations, regional constraints and highly diverse personalities thus combined to put Dorman Long in a largely reactive middle position politically, the USC's record was, in the main, one of active participation and partisanship for reform. Both economic interests and managerial social dispositions tended to put this firm in the vanguard on all the main issues: active promotion of tariffs, an intelligently applied cartelisation, regional amalgamations, central services on research, export selling and import buying, a strong reforming national federation evolving industry-wide policies, and sensitivity to both government policies and considerations of customer relations and public repute.

During the 1916–20 period Harry Steel set the tone for much that was to follow. He played a leading role in the NFISM, for instance helping it to recruit a chief executive from Whitehall, representing it before the Committee on Trusts, and forming a member of its top team in discussions with government. Steel took the lead in arguing for tariffs inside the NFISM, for example rejecting the view that free imports would be useful as a discipline for labour. He was reported as believing that 'the lesson learned by the men would be at far too great a cost to the nation as a whole'. On the other hand, there is no evidence that Steel accepted or even foresaw a degree of public surveillance as the likely price of protection. Indeed, he was probably a consistent free enterpriser who, it should be added, was also much occupied at this time in an angry battle with Whitehall over what he regarded as the government's 'monstrous' and 'abominable injustice' over its contributions towards the inflated costs of the new Appleby plant, a question which finally went to the Cabinet in 1919. Nor are there signs of any forward thinking on the issues of regional or nationwide co-operation. Here the running was left to John Craig with his prophetic private evidence to the Departmental Committee: 'If we are going to be purely individualistic, we can get no Government support and no other outside support. We have to choose between pure individualism and this combined method of dealing, and I think the advantages of the latter will outweigh those of the former.'[5]

During Albert Peech's chairmanship (1920–7) the USC began to move into the forefront on a wider range of issues. Harry Steel's policy of assiduous participation in the NFISM continued. Peech himself was involved in normal attendance at monthly executive meetings, work on sub-committees on labour relations and evidence to government committees, representations to Whitehall on matters like railway rates, and the Presidency in 1924–5. The USC's participation further increased when Maximilian Mannaberg became chairman of the NFISM's Fuel Economy Committee in 1923. Steel's advocacy of tariffs was maintained also: for example, in December 1925 we find Peech 'much engaged in the application for Safeguarding'. At the same time, however, the firm supported the more advanced Craig line on collective reforms inside the industry.

In 1922 Peech argued in the NFISM executive for 'closer co-operation and organisation between manufacturers to meet foreign competition...centralised selling for the development of the export markets...the possible application of similar methods to the Home Markets'; this was then very much a minority view. In 1927, when powerful interests jibbed at providing cost information to an official committee, Peech met its chairman, Lord Balfour, and reported back his conviction 'that it was in the best interests of the industry that the information should be given'. It should be added that Peech's industry-level enthusiasms were sometimes muddled. His 1925 presidential speech at the NFISM general meeting was characteristically confused. In 1924 his impetuosity in summoning a meeting of leading steelmasters to discuss a giant amalgamation project, involving Colvilles, Baldwins, Pearson and Knowles, Dorman Long and the USC, was subsequently disowned by the USC board, which forced him to withdraw. And his NFISM London involvements provided yet more ammunition for Mannaberg's climactic denunciations in late 1927 (see Chapter 5).[6]

It was during the Hilton–Benton Jones régime, particularly between 1930 and 1935, that the USC's pursuit of both industry-collective and public policy causes reached its highest pitch. Benton Jones brought to these issues convictions based partly on his long experience in coal, partly on temperament and ideals. In coal he played a leading role in organising co-operative marketing in the Midlands and by 1929 defined price pooling as doubly desirable: because 'economic necessities' made higher profits 'imperative' and because such a scheme, with its 'Rationalisation' content,'would tend to reduce the danger of political interference'. *Mutatis mutandis*, of course, both arguments could be applied to iron and steel. Interestingly, Benton Jones showed no great eagerness to exploit the public policy mantle of 'rationalisation' for the USC's commercial advantage. In early 1930, when others were already jostling for the BID Co's potential support with various (more or less reasonable) merger schemes, Benton Jones flatly told Montagu Norman at their first meeting that 'he was disposed to have nothing to do with any other steel concern for a couple of years'. He gave Norman the impression that 'his interest was coal, his duty steel', that he was 'surprised at being consulted by a Banker in London', that he was 'rather out of our current of thought on rationalisation' but intent instead on 'digging in' and 'local improvements'. This was not the behaviour of a politically agile corporate opportunist, although it appears to have won a certain regard from Norman.[7]

However, a man so immersed in concepts of duty soon became susceptible to the idea of the USC as a major contributor towards the political concessions which the iron and steel industry's collective interests seemed to require. After all, an interest in mergers, about which Benton Jones tended, in fact, to be sceptical, could be partially justified as a matter of political prudence on an issue 'much in the public eye'. He could see that the industry needed public goodwill both to avoid vague socialising threats under Labour in 1930–1 and to earn high tariff protection under the subsequent National Government. Benton Jones's collective inclinations also owed much to his concern for inter-company proprieties. For example, when the USC sought reduced trade association prices, reflecting its position as a below-average cost producer in the main, unilateral price-cutting and other high-pressure tactics were ruled out on principle. Benton Jones insisted that all matters concerning trade association behaviour should be referred to the board because they 'concerned the Company's good name as well as its immediate interests' – a revealing phrase. Later, he was to marry the USC's interest in cheap Midland iron ores with an abortive appeal for 'co-operation' and 'trust' in a giant consortium (see Chapter 6). Benton Jones could sometimes ride off on a political tangent, as when he publicly advocated import licensing in late 1931 while the NFISM's lobbying for tariffs was in full spate, an action which infuriated his industry colleagues. But in presiding over the USC's pursuit of a strong central body (described below) he showed a strong sense of 'economic necessity', desires for industrial teamwork and a public policy responsiveness at once pragmatic and dutiful.[8]

Robert Hilton's involvement was more concrete, forceful and radical. More than any of the other industry leaders he emerges as the arch-reformer of the early 1930s. In taking that stance he was partly influenced by the USC's sectional interests. For example, there was the fact that successive reorganisation schemes all envisaged that firm as an important fulcrum which might be added to and was unlikely to be reduced in size. There was the possibility of a few rationalising, publicly-approved acquisitions, about which Hilton could be quite robust, as when he hoped to dictate terms in a takeover of Partingtons in 1930. Like Benton Jones, though, he was no avid merger-monger, and both of the main USC acquisitions supported by the BID Co proved abortive: a commercially unpromising but publically worthy Cumberland rationalisation and the later, potentially more glamorous, idea of taking over the LSC (see Chapter 6). Again, there was the use of 'co-operation' as a lever, successfully between 1932 and 1935, less so thereafter, to cut in on Stewarts and Lloyds' grasp of cheap iron ores. There was also the desire to subjugate price-cutting competitors like Consett within a central cartel and a common front *vis-à-vis* a particularly troublesome group, the re-rollers.[9] But

Hilton's involvement went considerably beyond a purely corporate calculus. Like Benton Jones, he had an essentially moderate concept of the USC's advancement. He was also influenced by an impatient, tidy-minded newcomer's sense that the industry was in a mess and needed pulling together, an intellectual fascination with the ramifications of a comprehensive national organisation, a possible interest in the personal prospects offered by the new BISF, of which he became president in 1939, some understanding of Whitehall's point of view, and a consistent sense of the emergent moral compact between the industry and the state.

Already in 1929 we find Hilton taking the advanced view that standardised costs should be applied to the entire industry: 'the question is a national one.' In 1930, heavily involved with a few other industry leaders in helping Charles Bruce Gardner of the BID Co to develop a national reorganisation blueprint, Hilton went much further than his colleagues by arguing forcefully for a national selling company for heavy steel. About these discussions he wrote, not altogether disingenuously, to Gerald Steel: 'We are all trying to look at the problem on the broadest possible lines as a National one rather than playing for position in regard to our own particular areas.' By summer 1931 he was arguing that a new national body should also supervise the purchase of all raw materials, lock in its members for a definite ten years, apply a (flexible) system of production quotas, and supervise capacity expansions. No one else in the NFISM or the industry went so far, it seems. By autumn 1931 Hilton was bewailing the NFISM's purely lobbying tactics *vis-à-vis* the new National Government: 'Too little constructive work by the Industry which should be helpful to a Government Department in a time like the present.'[10]

From 1932 to 1935 Hilton's active membership of the National Committee gave him a platform for complaints that the industry was obsessed with protection and price maintenance and for ceaseless urgings in favour of 'a Central Controlling Body with real power' both for its own sake and as a fair *quid pro quo* for tariffs. He envisaged this body phasing out obsolete plants in an orderly manner, using a large, levy-financed fund: a mere 'squeeze' from market forces would, he thought, be hopeless because of bank interests in particular firms and continued irrational speculation. On the National Committee this view was shared by Sir Alfred Hurst of IDAC but, it seems, no one else. Hilton expressed anger that the committee's final reorganisation scheme of early 1934 involved still more watering-down of central powers over the trade associations and that the industry's 'reactionaries' should even then damage its 'reputation' by an equivocal reception of the scheme. During the summer 1934 debates on the role of the new federation's independent chairman, he argued that the latter should have 'all the power that he would have as Chairman and Managing Director of an individual company, leading his Board on matters of policy', plus even some duties 'outside the scope of the objectives of the Federation', a view actually more advanced than that put forward by IDAC and Whitehall. Interestingly, Hilton laced these opinions with private appeals to public policy considerations. Thus delays in setting up a central body would increase 'our difficulties' with 'the foreigner'; an independent chairman was needed to give 'some confidence to our customers that the Steel Industry is not out merely to exploit them'; and such a man should be acceptable to the steel industry 'but also to the Country in general'. By early 1935 Hilton was arguing yet another aspect of the emergent public–industry compact: if IDAC recommended increased tariffs the steel trade 'ought to make a gesture by reducing prices simultaneously with the imposition of the new duties'.[11]

In broad terms the USC officially and openly followed Hilton's lead. It took the initiative on ideas for central research in 1931, proposing a special levy for this purpose and offering to contribute more money if others would do likewise. On this issue Dorman Long supported the USC but Benjamin Talbot of South Durham opposed it. In 1932 the USC's top management agreed with Hilton that an 'impartial national body' should organise production quotas, 'direct general policy' and set up a fund to close redundant works, although it gibbed at the idea of a central fund also to finance extensions. In 1934 it was at one with him in taking a wide view of the independent chairman's powers and desirable characteristics. Its backing enabled Hilton to take a firm line at the climactic general meetings to consider the final scheme: 'As an industry they were absolutely committed to reorganisation...to turn down the Report simply meant going back on their word', and, later, 'If the Independent Chairman were not appointed, and the scheme not backed by a man of eminence and independence, it would not be brought home to the country that the industry was really in earnest' and 'the Scheme would not be a scheme at all.'[12]

Of course, it was one thing to take the lead in advocating radical reform, quite another to accept the disciplines which resulted once the new BISF under Sir Andrew Duncan got into its stride. There were bound to be difficulties, particularly over the system of voluntary price control started by the BISF and IDAC in summer 1935. After a brief honeymoon period of voluntary restraint up to spring 1936 the system ran into difficulties over conflicting policy criteria, basically as between 'cost-plus-fair profit', market demand and economic rationalisation. The resulting compromise was essentially a 'fair profit' policy qualified by supply–demand considerations and by resistance to extreme sectional pressures. There were large price increases in early 1937, which were maintained through the 1938 recession; then slight, though widely-contested reductions in late 1938; finally, further restraint in the shadow of a renewed upsurge of demand due to rearmament. This path-breaking experiment in peacetime voluntary price control produced some striking innovations, notably an extensive disclosure of company costs through the BISF to IDAC. However, it predictably opened up a Pandora's Box of problems: anomalies resulting from parallel free prices in supplying or purchasing sectors; forecasting difficulties over costs and demand; some protection of the inefficient; market fluctuations towards the end of the periods during which prices were fixed; evasion by recalcitrant groups; basic unsuitability to highly differentiated products.

The USC soon came up against Duncan on a number of issues mainly relating to price control. Its championship of West Cumberland led to protests against the BISF–IDAC line on steel rail prices in the home market in 1936 and 1937. It became increasingly worried about the methods used in ascertaining and defining costs for price control purposes. It was seriously annoyed with Duncan's line over coke prices, viewing the BISF's efforts to negotiate with a new coke manufacturers' association in 1937 as 'premature and ill-considered'. Internal anxieties about 'the Federation's methods and increasing powers' surfaced in a brusque exchange between Benton Jones and Duncan in early 1937 over the BISF's new system of monitoring investment projects. Benton Jones protested 'our unwillingness to surrender to the Federation our freedom to expand', to which Duncan replied that such 'freedom' was not in question, doubtless aware that the USC was unlikely to turn maverick despite such complaints (see below). By 1938 the price control régime was provoking further USC criticisms, by now largely related to the fixed-period system and the inflexibility of control during a period of weak demand when price reductions would have been preferred (although not by the Workington Branch).[13]

Nonetheless the USC essentially co-operated with the new régime. Its investment projects during the 1935–9 period fitted smoothly the newly-adopted BISF criteria in point of both careful pre-consultation and actual substance. Its disagreements were pursued constitutionally within the official structure. An internal decision to press for changes on a pricing issue 'but that in any case the Company must follow the decision of the majority' was typical. After the row on coke prices Hilton gave the BISF president a 'personal assurance that in other matters we have the greatest confidence in yourself and Sir Andrew Duncan and would hesitate a long time before opposing your wishes', adding to Duncan himself an expression of continued 'confidence in you as leader of the iron and steel industry'. Inside the USC the sales staff were imbued with the general ethos of 'not rocking the boat' and 'playing the [collective] game'. The USC continued to be amenable to political considerations affecting the industry's pricing: for example, 'In view of the imminence of the General Election it was felt undesirable to raise prices' (October 1935) and 'It is necessary to consider the importance of giving satisfaction to the consumer, for, as we have learned..., our industry appears particularly susceptible to public criticism' (May 1938). When an acute shortage of billets in 1938 led to an emergency scheme to supply the re-rollers, the USC claimed to be making 'sacrifices' whereas other firms involved were accused of being 'defaulters'. Its acceptance of the basic principles of co-operation was not in doubt, for example 'communal buying for the benefit of the trade as a whole' (Benton Jones in July 1939). Nor was Hilton's personal faith in the possibilities of the emergent 'Middle Way' seriously shaken. Thus in late 1936 he confessed privately: 'I do not think that our present control, a compromise though it may be, is bad for the Steel Trade in this country.' He would have preferred 'outside influences such as the consumers and labour...as some corrective to the policy of running the industry purely by those people who are engaged in the trade.' But the new system was preferable both to free competition (theoretically 'the most ideal way' but conducive to confusion and ruinous price-cutting) and to 'Government Control', with its attendant danger 'that the politician is inclined to give way to the labour element'.[14]

## Stewarts and Lloyds' separatism

The reader who has followed closely the characteristics and behaviour of Stewarts and Lloyds in earlier chapters will not be surprised that this firm pursued a separatist line on industrial and public policy issues, in marked contrast to the USC. Its economic position along the frontiers of the industry; Macdiarmid's assignment of complete primacy to corporate interests (Chapter 5); Stewarts and Lloyd's aggressiveness over its market power in tubes, its hold on Northamptonshire iron ores and its

invasion of iron and steel generally (Chapter 6); the firm's limitations on labour and local community issues (Chapter 8): all suggest the likelihood of major divergences between Stewarts and
Lloyds maximising entrepreneurialism and the compromises and
sensitivities inherent in the steel industry's movement towards a
middle way. So indeed it proved.

By the late 1920s Stewarts and Lloyds' management had
become habituated to a feeling that it belonged mainly to the tube
trade rather than to iron and steel, with a sense of mastery in
that trade and an adherence to long-term, rigorously sectional
goals.Although this much emerges clearly from the story so far,
two additional background points are pertinent. The first concerns Stewarts and Lloyds' traditional suspicion of collective
bodies generally. Thus in 1918, approached on the possible formation of an industrial council, the board 'did not consider that any
useful object would be served by such a body'. In 1921 it had
doubts about continued membership of the Federation of British
Industries (FBI). In 1925 it was annoyed with the FBI's
apparently compromising attitude over the Trade Disputes Act
and, invited to a meeting, insisted on previous undertakings
which were, not surprisingly, refused. Although Stewarts and
Lloyds was already a sizeable iron and steel producer, and hence
a member of the NFISM and the relevant iron and steel industry
trade associations, it shows no sign of participating in their counsels. As late as 1931 its steelworks committee pronounced that
the research levy proposal (see above) was 'inopportune'. Even in
the British Tube Association Stewarts and Lloyds' role was
apparently limited to normal cartel operations and did not extend
to research, statistics, export promotion and the like, although it
should be added that probably very few others in that atomistic
trade would be favourably disposed to such pursuits either.
Whether this policy of 'keeping oneself to oneself' was meritor-
iously businesslike or selfishly narrow raises wider issues. The
fact is, it made for habits which would be hard to break.[15]

The second point applies particularly to Allan Macdiarmid's
style and concerns his tendency to be an exceptionally hard bargainer. A zero-sum, 'winner-loser' psychology and the need to feel
personally and corporately victorious were evident to the point
where quibbling and tricky manoeuvring sometimes seemed irre-
sistible to him. Macdiarmid's sharp manoeuvrings with the BID
Co in 1930–2 and his abrupt break-up of the liaison with the USC
in 1936 have already been referred to (Chapter 6). Similar patterns were evident in the long-protracted negotiations with John
Craig of Colvilles in 1933. Stewarts and Lloyds' asking price for
its Scottish plate business, to be transferred to Colvilles as part of

the deal, was inordinately high; it yielded on the idea of a cash payment but later insisted on one; it suddenly pushed hard for entirely fresh clauses at a late stage in the discussions, further restricting its obligations to take plate from Colvilles; and whereas Craig referred to the Scottish and 'national interest' implications of the agreement, these were never mentioned by Stewarts and Lloyds. These quibbles must be seen in the light of the fact that Colvilles was effectively relieving Stewarts and Lloyds of a large part of the social embarrassment of its historic withdrawal from iron and steel in Scotland.[16] A similar streak emerged in the long negotiations with John James of LSC, starting in early 1937, where once again Macdiarmid was personally involved. Although James was also a hard bargainer, it was Stewarts and Lloyds' refusal to accept a provision that the LSC's auditors should have detailed cost information which led to a long delay and a resort to arbitration. The Company Secretary himself privately noted: 'It was here I thought we were wrong.' Even in 1939 arguments persisted on petty matters concerning the agreement. Clearly, such bargaining patterns, not evident in the records of either Dorman Long or the USC, might be hard to reconcile with a co-operative mode.[17]

Part of the problem was that Stewarts and Lloyds' intensified dominance of the tube trade involved strong elements of dissimulation. The intensity of international competition particularly in 1929–32 and again in 1935–6; public suspicions of monopoly; the huge increase in market power implied by the firm's whole strategy: all this implied secrecies and subterfuges. Thus the firm was less than forthcoming about its market share in tubes in confidential talks with, for example, Montagu Norman. Its attitude towards the tail of remaining 'independent' tube firms was equivocal. Their presence avoided the appearance of outright monopoly, they could be used as a partial cover for price increases, their inefficiency set Stewarts and Lloyds' own advances in rosy high relief and yet, inevitably, its policy was 'opposed to assisting [them] to reduce costs'. The Scottish Tubes Company was taken over and quickly run down but its name and commercial representation were maintained, implying a pretence of continued distinctness. The agreements with a firm like the Wellington Tube Works bound both firms to keep their close co-operation secret from buyers 'to safeguard against any prejudice' and 'to do everything possible so as to secure that their travellers, representatives and agents do not disclose the existence of such working arrangements'.[18] Of course, such dissimulations were hardly unique to Stewarts and Lloyds, although cartel practices in iron and steel were generally more open. But the point once again is that they contributed to a state of mind which marched uneasily with the elements of trust and wider give-and-take involved in the middle way.

Stewarts and Lloyds' first direct contact with public policies, its financial sponsorship by the BID Co, finally agreed in autumn 1932 (see Chapter 6), did little to reduce these ambiguities. If anything, it increased them. Theoretically, the fact that BID Co (and by implication, Whitehall) supported the firm's tube monopoly and its strategic potential for ultra-cheap steel-making, on top of the recently acquired benefits of tariff protection (all offering prospects of greatly increased power and profit for the firm), could have been used to exact some undertakings on market behaviour. ICI had been set up under public auspices in 1926 with a clear view of certain public duties, and the newly-formed IDAC was already pursuing the notion of a tariff-reorganisation compact in iron and steel. Yet somehow it was assumed that Stewarts and Lloyds' rationalisation plan was publicly virtuous enough to require no accompanying strings save a scaling-down of the scheme (which, in fact, partially vitiated it) and the liaison agreement with the USC (which merely represented a partial buy-off of another sectional interest which cost Stewarts and Lloyds little). No conditions on long-term behaviour within iron and steel, or pricing or even continuous disclosure, were stipulated.

Of course, this one-sided situation partly reflected the sheer confusion of public and semi-public bodies and of public policies at this time. As sponsor, the BID Co was confined to a rationalisation-supporting role and, anyway, was not alert on monopoly problems. The IDAC was still feeling its way and was not directly involved. Whitehall was only indirectly concerned, given the Bank of England's substantial independence, and in any case was hardly very clear itself on industrial policy dilemmas between 'rationalisation' and 'anti-trust'. That Stewarts and Lloyds had already sought to exploit these various cross-currents and uncertainties was hardly surprising. Certainly, it benefited from them. Its substantive victory would have mattered less if the firm had been publicly-minded in a spontaneous sense rather than just politically adroit, but it was not. There is no evidence that Macdiarmid and his colleagues, convinced as they apparently were that Corby itself would be super-abundantly patriotic, had any other concept of duties to the industry or public policy which might restrict their market behaviour or profit-taking.

It was hoped that Stewarts and Lloyds would contribute to reform discussions in iron and steel, reflecting its leading position. In his influential report of 1930 Charles Bruce Gardner of BID Co, had suggested that, although the firm's tube concentration made for some genuine separateness, it should nonetheless be incorporated within the iron and steel industry's machinery. Between 1932 and 1935 Stewarts and Lloyds did make some limited efforts along this line. Macdiarmid joined the National Committee and two of its Regional Committees, and in 1934 he served on the small committee on the proposed new BISF's chairmanship. But the participation was restricted and equivocal. Macdiarmid consistently held a poor view of the industry's leaders, doubtless partly because of feelings of intellectual superiority, partly also because of an outsider's lack of understanding of the genuine constraints. His views were clearly out of sympathy with the inefficient backwoodsmen who obstructed change but also with the multilateral, collectivist solutions advocated by someone like Hilton. The firm's own isolated foray along that line was narrowly sectionalist. In spring 1933 it joined with some other BTA firms in a proposal to set up a tight club of tube firms with restrictive agreements both among themselves and with iron and steel firms, designed to safeguard demarcation, control new entry and prevent undesirable newcomers getting hold of raw materials, particularly imports – all under the umbrella of a national reorganisation. This concentration on demarcation and entry control in its own sector was typical. Not until the early 1940s did Macdiarmid apply his mind constructively to the problems of steel industry economic organisation as a whole.[19]

When it became clear in 1935 that the new BISF under Duncan would be highly active Macdiarmid got distinctly cold feet. It must be said that there was a lack of sympathy between the two men, although whether that preceded or followed Stewarts and Lloyds' growing deviationism is not clear. In summer 1935 IDAC recommended continued tariff protection for the tube industry, praising its increased efficiency. The year brought bumper profits but also renewed international competition for Stewarts and Lloyds. In early 1936 Macdiarmid moved over to an aggressive line, privately putting the firm on a domestic war footing in a key paper to the board, declaring in advance in that paper his extreme reluctance to submit to collective BISF disciplines and, by April, ending the USC liaison agreement (see Chapter 6). Thereafter the separatism intensified. Even by 1939 the tube makers remained unaffiliated to the BISF. In 1937, in connection with the latter's collective import buying scheme, Stewarts and Lloyds caused great annoyance by refusing to pay levies on steel ingots it made into tubes.[20]

More striking was Stewarts and Lloyds' further expansion of steel-making at Corby, starting in 1936. Arguably, the firm kept within the letter of the 1932 understanding but, perhaps, only just within it, by including a partnership element, this time with the LSC. Also, there were strong arguments in favour of the move in terms of both scale economies and international competition. But elsewhere in the industry it was viewed as provocative and irresponsible just when the BISF was attempting some informal monitoring of investment projects in order to forestall feverish speculation and uneconomic duplication of plant in an incipient investment boom. Stewarts and Lloyds did not match its further incursions into steel by giving up the corollary 'demarcation' principle of no iron and steel firm entering tubes. There was no serious evidence of anyone actually doing this, and they would have incurred BISF's severe displeasure if they had. Stewarts and Lloyds also continued to sidestep proposals for a new co-operative development of the Northamptonshire iron ores. By March 1937 Duncan and Macdiarmid had met to discuss their differences but apparently to little effect. It is worth quoting Duncan's views on the deterioration, expressed at BID Co meetings:

> Mr. Macdiarmid had become more and more rapacious and, whereas originally his thought was for the protection of Stewarts and Lloyds against inroads into their tube business, he was now regarding himself as a steelmaker. (October 1936)
>
> Sir Andrew referred to his efforts to keep steelmakers out of tubes ... and to confine Stewarts and Lloyds to tubes, the only sale of steel being that required to round off their production. (November 1936)
>
> Stewarts and Lloyds need to be safeguarded against a steel industry invasion into 'tubes' but must not themselves invade 'steel'. But since they own ore deposits in Northamptonshire which should be available for a much wider use than their own tube production, as owners of ore (as distinct from tube manufacturers) they could properly be associated in a joint enterprise with the other two principal steelmakers of the Midland area, viz: US and LS. (November 1936, letter to Montagu Norman)
>
> Sir Andrew discussed the attempt which he thought Stewarts and Lloyds were making to 'express themselves' as steelmakers, an attitude which he deprecated without some collaboration or agreement with other makers. (September 1937)

When directly challenged by Montagu Norman in September 1937 on the iron ore consortium idea, Allan Macdiarmid's response was cloudily helpful but in effect stonewalling. Norman reported 'goodwill on principle, subject to reservations as to method'. Macdiarmid took the opportunity to let off steam to Norman about the whole new BISF–IDAC régime: 'failure of planning', 'modernisation of plant swept away by protection and prosperity', 'unwillingness to use Brassert, a genius who was now swallowed up by Germany.'[21]

Stewarts and Lloyds' other major deviation was over pricing policy. Here the firm was able to take advantage of the fact that the new price control arrangements were voluntary. Although some small trade associations also resisted the system quite successfully, notably the Hematite Pig Iron Association and the foundry pig iron producers (both representing mainly small, atomistic, high-cost producers), Stewarts and Lloyds was the only large firm consistently to run rings round the scheme. It was able to use its position 'half in, half outside' the new BISF and to exploit the vulnerability of a form of social control based on 'voluntarism'.

Stewarts and Lloyds raised prices in April 1936, September 1936, April 1937 and late summer 1937. In official quarters there were doubts whether these increases were justified in view of the firm's undertaking not to raise prices when tariffs were increased. The 1936 price increases left IDAC 'somewhat disturbed', partly because they were rushed through. One rise occurred 'without notification', the other 'while discussions were proceeding'. Nonetheless IDAC reluctantly agreed to them. Stewarts and Lloyds' disposition to shelter behind the need of inefficient small firms for an increase was hardly questioned. There was substantial acceptance of the firm's arguments as to increased costs, the need to cross-subsidise exports ('Stewarts and Lloyds should be in a position to hold their own'), and 'some extra rewards, for a period at least, for the risk taken...in large schemes of modernisation'.

By August 1937, though, IDAC's suspicions had sharpened. With the studied under-statement typical of this semi-official body, its secretary complained that the previous April's increase, 'an appreciable addition', had been notified only 'informally'. Increased input costs had been 'very largely offset by the economies resulting from greater production', very high export prices were now 'admittedly profitable', and 'there appeared to be a strong *prima facie* case against proposals' for yet another increase. To Stewarts and Lloyds' plea that 'the increased price was essential to give [the independent small firms] a reasonable profit' IDAC replied with some asperity: 'The price ought not...to be governed by the costs of the Midland makers, who it was clear were only existing on sufferance by permission of Stewarts and Lloyds and must not be used by them as a stalking horse to get better prices.' The Secretary suggested, 'Clearly they are not adhering to their undertaking', although some mitigating arguments were recognised. But 'It became clear in the discussions that while they (Stewarts and Lloyds) attached considerable importance to securing the Committee's goodwill they were not prepared to defer the increase...and had in fact already put it into operation in renewing some long-term contracts.'

Once again IDAC had no alternative but to acquiesce with no chance of redress, or of publishing its disagreement with the firm, or even of exacting more information. All it could do was piously state to Stewarts and Lloyds 'that the profits on the Corby position will be abnormally high and that in the long-run they will expect some part of these to be passed on to the consumer'.[22]


## Conclusions on social action

We have now considered virtually the whole field of social action. A brief examination of labour and local community issues in the last chapter led to the inference that the USC was well ahead of Stewarts and Lloyds, overall, with regard to issues of worker advancement, employment and redundancy, local community responsiveness and philanthropy (the evidence on Dorman Long was less clear). The present chapter has considered the more complex question of public policies for the industry. It has shown that this area posed significant problems of managerial choice and that the firms' responses fell into a definable pattern: Dorman Long's passivity, the USC's co-operativeness, Stewarts and Lloyds' separatism. It is necessary to conclude this section of the book by relating these findings briefly to the central contentions about growth–efficiency–social action trade-offs, managerial specialisation and corporate development phases. Three key questions here concern the comparability of social action between firms (convergent or multiple?), its derivation (exogenous or endogenous?) and its importance within business policy (as an objective or as a constraint?).

The question of inter-firm comparability can be dealt with quickly. Social action was defined in Chapter 1 as the voluntary pursuit of publicly-approved activities outside or beyond the firm's own growth and efficiency as sectionally conceived. These activities emerge as so diverse that firms' social action cannot easily be categorised and ranked as a whole. In the case of both efficiency and growth a variety of processes can be grouped together, so that a firm's overall showing can be classified defensibly as, say, 'good', 'middling' or 'poor', relative to comparable firms. But in the case of social action this is not so. A firm may achieve different positions in the league relative to, for instance, local community issues, collective industrial action or responsiveness to government, and no effort to weight these positions in order to arrive at a composite measure is likely to be commonly acceptable. It is interesting that the USC did, in fact, come out fairly clearly in front on all the tests, Stewarts and Lloyds behind. But such a clear overall ranking was not necessarily typical. It seems more reasonable to view social action as a wide field with several corners in some of which a firm might excel or fail but not others.

The question of derivation is more difficult. Does the evidence support the contention in Chapter 1 that this managerial inclination, where it existed, was no mere derivative of growth or efficiency-seeking nor, for that matter, reducible to would-be profit maximising, but something with its own largely independent roots?

To a considerable degree it appears that social action, at least on industrial policy and relations with government, interlinked with sectional economic interests. For example, as large heavy iron and steel firms all through, both Dorman Long and the USC might be expected to support cartelisation of that trade and its environs, and a strengthening of its central organisation particularly *vis-à-vis* foreign competition. By contrast, Stewarts and Lloyds' tube specialisation placed it largely outside the iron and steel industry proper at least before 1932. Therefore, once both tariffs and Corby were assured, its virtually single-handed dominance of the tube trade required little help from collective action. Even Stewarts and Lloyds' increasing importance in iron and steel after 1932 involved ambivalence. Although cartelisation would be helpful to it, the resulting collective disciplines might be unwelcome in so far as the firm sought to exploit its position as a newcomer, a cheap producer and the virtual king of the cheap Northamptonshire iron ores. Again, economic considerations were relevant on particular collective issues, notably the firms' varying dependences on imported materials or domestic scrap supplies, the relative importance of high, medium or low cost plants in their output mix, and their varying interests in further mergers or in preserving the corporate *status quo*.

However, such sectional economic interests were emphatically not the sole determinants of responses to collectivism and politicisation. Whether a firm clearly perceived its economic interests; whether these were clear-cut anyway, particularly in view of short- versus long-run conflicts; how politically active the firm might be in helping forward a collective economic interest at regional or national levels; how far it would yield, in practice, to the corresponding collective disciplines, let alone the publicly-enjoined restraints: on all of these matters great scope for variation existed on which 'economic interests' would be only the broadest of guides. Correspondingly, other influences were also at work: personal likes and dislikes (as in Macdiarmid's relationships with both Benton Jones and Duncan); corporate traditions on collective action; interests in public repute or personal advancement through collective action (probably a factor for Hilton); contacts with public policymakers (again evident in Hilton's case); questions of temperament, 'clubability', etc., affecting whether an industrialist was regarded as a team player or a maverick; and, not least, management's ideological attitudes (including, for example, Sir Arthur Dorman's nationalism, Sir Hugh Bell's classical liberalism, Macdiarmid's implicit neo-classicism, and the co-operativist, middle way concepts of Benton Jones and Hilton). As for the labour and local community issues, I have suggested that childhood influences, civic affiliations, the extent of previous links with wider social groups and, again, ideology played a part. To reduce this wide range of social, personal and ideological factors to epiphenomena of economic forces, growth–efficiency seeking or profit, is an extreme form of reductionism or monism, both philosophically and empirically unacceptable.

Finally, there is the related question of the status of social action in the hierarchies of managerial objectives. Why not consider social action as a more or less powerful 'constraint' rather than an objective standing in its own right alongside growth and efficiency?

It is obvious that even strong social action pursuits claimed less of the foreground of managerial time and attention, overall, than growth or efficiency. But, first, the intensity of the relevant sentiments and convictions must be taken into account. Even a tiny call on resources for, say, philanthropy, retention of an elderly employee or public justification could evoke deep psychological drives or ethical ideals. The fact that social action lay outside the obvious constraints of obedience to the law meant that its interpretation varied just as its frontiers were subject to change, not least by business firms themselves. Like growth and efficiency it was hazy as well as a 'flying goal', to use Schumpeter's phrase. Some social action pursuits even approached the category of quantifiable maximands. Dorman Long in the 1920s doubtless thought of itself, quite reasonably, as maximising its donations to local charity, just as the USC in the 1930s undoubtedly sought to maximise job provision in West Cumberland, both, of course, subject to profit constraints. Within this context of social action being variable and dynamic there was the question of whether standards of employee welfare should be 'as good as other progressive firms'' or 'the best', whether social action depended on spontaneous conviction or outside pressure (waiting to be pushed), whether whole policies like joint consultation or workers' pensions should actually be pioneered in the industry and whether management should also pioneer by advocating advanced forms of collective action, as Hilton and the USC repeatedly did.

Obviously, as we have seen, the answers varied enormously in practice as well as theory. Many of the responses, particularly in Stewarts and Lloyds' case, can fairly be classified as reactive or imitative, 'constrained', 'satisficing' or even 'minimising'. But for many other responses, particularly in the USC's case, such categorisation would be artificial and misleading. Overall, it seems reasonable to see social action as (a) an area largely *sui generis*, not an economic derivative; (b) a field of choice widely viewed as morally fraught by managers as well as public opinion; and (c) a fully-fledged 'objective' at least in certain managerial phases and firms. But this whole field certainly requires more study.

## Notes

1   D/ACS. D/AGMs. Departmental Committee of Enquiry into the Iron and Steel Trades after the War, Evidence of North East Coast Steelmakers' Association, January 1917, in PRO/BT 55/40. For other evidence see also BT/ 55/38 and 39. F/EC, 19.10.22 and generally through 1923 and 1924, and also Parliamentary and General Purposes Committee. J. C. Carr and W. Taplin, *History of the British Steel Industry* (1962), p. 535. North East Coast Steelmakers (260/1/1/2/13/1) and *ditto*, Minutes (210(d)15/22). D/DM, 16.3.26, 13.4.26, 20.5.26, 22.7.26, 14.9.26, 12.10.26, 17.7.28, 18.9.28, 13.11.28.

2   Sir Hugh's high public profile reflected wider business activities (including directorships of the Yorkshire Insurance Company, Brunner Mond and the North Eastern Railway), involvements in local affairs over several decades (Middlesbrough town council, twice mayor, Lord Lieutenant of the North Riding, etc.), and his contributions as a politician and public figure at national levels (the 1916 Departmental Committee, the Science Museum, Imperial College, numerous speeches as a free trader and liberal individualist). *Who was Who* (1929–40), obituary, *Northern Echo*, 30.6.31. Many of Bell's opinions emerge from his annual speeches as chairman of Bell Brothers up to 1921: see DL, Bell Bros., Reports, Accounts and OGM Proceedings (1066/24/1). See also his contributions in S. J. Chapman (ed.), *Labour and Capital after the War* (1918) and *Contemporary Review*, Dec. 1925 and Feb. 1926. In his later years Bell's anti-*dirigisme* included opposition to electricity nationalisation and the 1930 Coal Marketing Act.

3   Much of the National Committee's tortuous saga can be extracted from NFISM Film 39; Executive Committee Minutes between 1932 and 1934; and Minutes of general meetings during this period. The IDAC Papers in PRO/BT/10 are important: see particularly CP 265/32, 88/33, 263/33, 17/34, 24/34, 146/34, 168/34. Useful references, including some critical comments,

will be found in BE: SMT 9/3 and 4, SMT 3/89 and 90. See also IDAC Reports, Cmd. 4117, 4181, 4589; H. J. Hutchinson, *Tariff-making and Industrial Reconstruction* (1965), pp. 131–40; and Carr and Taplin, op. cit., pp. 495–502. With his departure from D L (see Chapters 6 and 7) Mitchell also disappeared permanently from the national stage. He joined a Dutch reclamation company but was later given a position within the BISF during the Second World War. Interviews with retired DL officials.

4 DL, Managing Director's Monthly Reports (1066/16/11). L. Ennis to Sir C. Wright, 27.6.35. IDAC Papers, PRO/BT 10, CP 53/37. D/DM, 18.5.34, 28.6.35, 13.9.35, 27.9.35, 20.1.37, 12.3.37, 8.4.37, 25.6.37, 28.7.37, 21.9.37, 29.10.37. Interviews with retired DL officials. As President of the BISF, 1945–52, Ellis Hunter became a leading architect of its policies during the postwar years.

5 F/EC, 3.12.18, 19.12.18, 20.3.19, 10.4.19, 30.4.19, 22.5.19, 19.6.19, 17.7.19, 24.7.19, 18.12.19, 21.10.20. F/GM, 17.7.19, 25.3.20. Iron and Steel Industries Committee, PRO/BT 55/39, Minutes of Evidence, 1916, Scottish Steel Makers' Association, Mr J. Craig.

6 F/EC, particularly 15.6.22, 19.10.22, 16.7.25, 17.3.27, 21.4.27; Parliamentary and General Purposes Committee, 14.7.26, 16.2.27, 16.3.27; F/GM, 21.5.25, presidential speech by A. O. Peech. U/DM, 5.8.24, 6.10.25, 8.12.25.

7 U/DM, 12.2.29, and for later activities on coal reorganisation schemes U/DM, 24.2.32, 22.6.32, 22.2.33, 26.4.33, 26.7.33, 22.11.33, 20.12.33, 28.11.34, 21.12.34, 24.7.35, 23.10.35. CI, notes on Rother Vale Collieries, 15.7.31. BE: SMT 2/147, memorandum on Norman–Benton Jones meeting, 19.2.30. Norman described Benton Jones as 'extremely friendly' (*ibid.*) and 'a quiet man better informed on coal than on iron and steel' (letter to Sir H. Goschen, 20.2.30).

8 For Benton Jones's general scepticism on mergers see Chapter 5. U/CC, 4.8.32. Chairman's Finance Policy Memorandum 1933 (006/10/1). NFISM, Film 3, correspondence involving E. J. George, M. S. Birkett, Sir W. Larke and W. Benton Jones, January 1932.

9 BE: C. Bruce Gardner, Report on the structure of the iron and steel industry of Great Britain, Dec. 1930. For Montagu Norman the USC was 'a keystone...not entirely similar to Beardmores for although it contained a permanent element it might either be added to or taken away from': BE: SMT 9/1, 23.6.30. Hilton's emphasis was rather different: 'I believe we shall get assistance from the City in finding money for plant extensions, but it will, of course, necessitate the shutting down of a number of works; none of ours I hope will be affected in this way', U/MD, 6.8.30. U/MD, R. S. Hilton to W. Benton Jones, 24.9.30. For the abortive West Cumberland merger efforts see U/DM, 11.11.30, 27.11.30, 24.2.31, 9.4.31, 27.5.31, 30.6.31, 28.7.31, 10.11.31, 8.12.31, 27.1.32. Also, BE: SMT 3/73, memorandum by W. Benton Jones, 10.7.30; SMT 3/18, interview with Benton Jones and Hilton, 27.11.31; and more particularly SMT 7/1 and SMT 3/238. For the price-cutter subjugation motive see, for example, U/MD, correspondence between R. S. Hilton, E. J. George and C. Mitchell, Aug. 1932, also 12.1.33, 9.2.33, 11.2.33, 13.2.33, 8.7.33.

10 BE: SMT 3/18; SMT 3/74; SMT 7/2; and SMT 7/1, especially R. S. Hilton to C. Bruce Gardner, 6.9.30, 6.11.30, 26.9.31. U/MD, R. S. Hilton to C. U. Peat, 31.7.29, to G. Steel, 1.10.30, to Sir W. Larke *et al.*, 14.11.30, to A. Dorman, 18.11.30, to Sir J. Beale, 18.11.30, to Sir W. Larke, 20.11.30, 23.1.31, 4.7.31, 4.9.31, 22.9.31 and 30.11.31, to P. Lindsey, 30.11.31, to Sir W. Larke, 16.12.31, 24.12.31, to J. I. Piggott, 2.1.32, to G.S. McLay, 7.1.32, to Sir A. Duckham, 26.1.32, to Sir P. Rylands, 25.2.32.

11 U/MD, R. S. Hilton to G. S. McLay, 29.4.32, to Sir W. Larke, 23.5.32, to E. J. George, 7.10.32, to Sir A. Hurst, 23.11.32 and 27.1.33, to Sir W. Larke,

12.4.33 and 29.11.33, further letters to Hurst, Jan. 1934, to Sir George May, 25.1.34, to Sir W. Larke, 17.3.34, to A. C. Macdiarmid, 29.3.34, to Sir W. Larke, 29.3.34, to C. Bruce Gardner, 9.4.34, to Captain A. H. Reed, 1.5.34, to Sir A. Hurst, 23.6.34, to M. S. Birkett, 25.6.34, to R. A. Dyson, 26.6.34, to Sir A. Hurst, 19.1.35 and 14.2.35.

12   U/DM, 8.9.31, 26.9.32, 27.2.35, 24.4.35. U/CC, 19.11.30, 19.1.32, 5.4.32, 25.10.32, 18.11.32, 12.7.33, 11.4.34, 18.5.34. F/EC, 21.4.32, 18.5.33, 18.1.34, 3.5.34, 10.4.35, 10.7.35. F/GM, 17.9.31, 22.2.34, 19.4.34.

13   For the general development of voluntary price control see IDAC Papers, PRO/BT 10/13–29. See also Carr and Taplin, op. cit., pp. 558–66, 572–7. U/DM, 24.3.37, 27.10.37, 14.12.37, 16.12.38, 25.1.39, 22.2.39. U/CC, 9.9.36, 9.12.36, 10.2.37, 19.5.37, 11.8.37, 13.10.37, 8.12.37, 8.6.38, 10.8.38, 11.1.39, 9.2.39, 14.6.39. U/MD, R. S. Hilton to R. Crichton, 5.12.37.

14   F/EC, 21.1.37, BISF, Film 4, Expansion Committee, 14.1.37. U/CC, 14.7.37, 8.9.37, 10.11.37, 12.1.38, 9.2.38, 12.7.39. DL, North East Coast Steelmakers (260/1/1/2/13/1), comments by G. A. Chicken of USC 19.5.38. Author's interviews with retired USC officials. U/MD, R. S. Hilton to Sir C. Wright, 28.6.35, to L. D. Whitehead, 21.10.35, to W. Benton Jones, 23.12.35, to H. A. Brassert, 27.11.36, to Sir A. Duncan, 4.12.36 and 24.12.36, to Sir C. Wright, 10.12.37, to Sir A. Duncan, 22.12.37.

15   S/BM, 7.3.18, 24.10.18, 27.10.21. S/GPC, 20.7.21, 24.9.25, 22.10.25, 19.11.25, 17.12.25. Steelworks Committee, Minute Book 4, 7.9.31, in SL (1791/1/14). There is no hint of collective economic reformism in S/OC, where most routine tube trade association matters were reviewed, nor in the occasional surviving notes of trade association deliberations, e.g. Tube Makers' Association, papers, 1879–1931 (1791/1/22); Minutes of meeting of Large Tube Makers, 23.1.29; Evidence of T. C. Stewart on behalf of British Tube Association in Iron and Steel Industries Committee, 1916 (PRO/BT 55/39); and BTA reactions to National Committee's first reorganisation scheme, 30.5.33 (NFISM, Film 39). But conditions and customs in the tube trade would have been unfavourable.

16   Colvilles, Mr Craig–Stewarts and Lloyds (1283/1/7), correspondence in 1933 involving J. Craig, A. C. Macdiarmid, A. McCance, J. Mitchell and W. Stewart. David Colville and Sons, Minute Book (Glasgow 1283/1/10), 11.11.32, 19.12.32, further references through 1933, 9.3.34. P. Payne, *Colvilles and the Scottish Steel Industry* (1979), pp. 206–8. For S and L's stance in earlier discussions on Clydeside mergers, see Payne, op. cit., pp. 182, 199–201. Macdiarmid had been 'infuriated by the arguments of the Scottish makers', S and L had imposed 'totally unacceptable conditions' and shown 'intransigence'.

17   S and L, Lancashire Steel Corporation (002/2/1), includes notes of meetings and correspondence involving J. E. James of LSC and, on S and L's side, A. C. Macdiarmid, J. Menzies-Wilson, F. Scopes and F. McClure. S and L's decision-making rationale was expressed with characteristic clarity in a draft for report to board, 27.7.37. The lawyers were involved by early 1938. For the reservations inside S and L see F. S. Scopes to C. C. Smith, 25.3.38, including pencilled comment in margin by Smith. See also LSC, Notes on meetings, 1937–9 (002/2/4).

18   BE: SMT 2/148, secret aide-mémoire for A. C. Macdiarmid in talk with Montagu Norman, 21.1.30. But see also C. Bruce Gardner, Report, Dec. 1930, op cit. S/GPC, 13.6.19 and report to board, 23.7.19, 1.6.31. Economy Commission (1791/1/21), Report from G. F. Satow, 9.9.32. Agreements between S and L and Wellington Tube Works (1791/1/12). For attitudes to UK competitors see also Chapter 6. For their uses to S and L in defending price increases see below and IDAC papers, PRO/BT 10/19, CP 197/36 and CP 174/37.

19   BE: C. Bruce Gardner, Report, Dec. 1930, op. cit. F/EC, 2.8.34, 15.11.34. On the independent chairmanship issue in 1934 Macdiarmid appears to have aligned himself with the USC: see Consett, Reorganisation of iron and steel industry (01999), sub-committee (of NFISM) on 'Independent Chairman'. See also F. Scopes, *The Development of Corby Works* (1968), p. 229. But the NFISM papers confirm S and L's generally low profile in both private and general discussions on reorganisation, 1932–5. NFISM, Film 39, comments by BTA (S and L, Tube Investments and six other firms) on the National Committee's first reorganisation scheme, 30.5.33. For Macdiarmid's post-1939 views see D. L. Burn, *The Steel Industry 1939–1959* (1961), pp. 59–60, 67–9, and Macdiarmid himself in *The Times*, 18.5.45.

20   IDAC papers, PRO/BT 10/15, CP. 186/35. S/BM, 2.3.37. D. L. Burn, op. cit., pp. 57–8. Burn defends Macdiarmid's general position: 'He found it hard to believe, at this time, that a committee of steel firms would vet each other's plans in a disinterested way, or that a body of disinterested people could vet them competently.'

21   S/BM, 29.10.35, 15.6.36, 30.6.36, 28.7.36, 2.3.37. Letter to shareholders, 30.9.36. Lancashire Steel Corporation (002/2/4). Scopes, op. cit., p. 235. BiSF, Film 4, correspondence between L. D. Whitehead, G. H. Latham and Sir A. Duncan, Oct. 1936. BE: SMT 9/6, 5.10.36, 19.10.36, 2.11.36, Sir A. Duncan to the Governor, 10.11.36; SMT 9/7, 6.9.37, 1.11.37. SMT 2/148, Memorandum on meeting, Montagu Norman and A. C. Macdiarmid, 29.9.37. Norman said Macdiarmid objected to 'the Midland Group' because 'amalgamation was unthinkable' and 'Firth (of Richard Thomas) could not be trusted', adding: 'it was too clear that he would like the first step to be his purchase of LSC. I ignored this hint deliberately and openly. Towards the end he *seemed to admit that he was in favour of the Regional Group on national grounds* provided it would not cut S. and L's throat and would represent less than an amalgamation', (my italics). 'We decided that the first thing was for the parties to be agreed in principle...and he was inclined, with some hesitation, to write to Benton Jones and to see how far these two alone could meet and get towards a solution.' In view of previous events (see Chapter 6), Macdiarmid can hardly have been serious.

22   IDAC Papers, PRO/BT 10/19, CP/197/37; and BT 10/23, CP 161/37.

# 10 Conclusions

This book has taken us through massive changes in the economic context of British iron and steel: the excitements of the First World War, the long astringencies of the depression years, the revival of the 1930s. Politically, it has spanned wartime government controls, the last long phase of *laissez-faire* in the 1920s, the move by the 1930s towards tariffs and informal public control. Our story has brought us through large numbers of strategic decisions under as many as eight different managerial régimes in the three large companies.

On any criteria, certain managerial achievements covered in this book emerge as outstanding. Such were Stewarts and Lloyds' dazzling growth, both nationally and internationally, between 1928 and 1939, Harry Steel's break-neck formation of the USC (1916–20) and the USC's classic turn-round between 1928 and 1932. To these must be added a few other specialised achievements, notably Dorman Long's overseas ventures in constructional work, Stewarts and Lloyds' pioneering on information systems in the early 1920s, and the USC's major contributions to social policy and to the industry's political mutations in the 1930s. But since a persistent priority has been to avoid a triumphalist, Whig species of management history, the themes of error, conflict, pain and ambivalence have also loomed large.

One thing the book has not done is to chronicle each firm with the circumstantial detail usual in single company histories. Consequently, a number of interesting issues have been excluded. The detailed treatment of, for example, pre-1914 developments and technological or financial aspects falls to others. Instead, the aim has been to shed light on a few central themes. Technical complexity has been avoided in order to concentrate on the derivation, formulation and implementation of business policies. Within that context, already representing a simplifying model of complex historical realities, certain hypotheses have been pursued relating to (1) degrees of rationality and error in decision-making, (2) trade-offs between growth, efficiency and social action, and (3) managerial specialisation and phases in corporate development.

Before considering how far the study supports the specific ideas advanced in the first two chapters, it is necessary to ask what lessons emerge about the general nature of business policies.

One theme relates to the indivisibility and integratedness of business policies, also their deep cultural roots. Long-term plans, regular decisions and responses to crisis all bear witness to their synoptic character. Within the crucible of top management decision-making all subject and functional divisions within the firm are continually dissolved and re-combined in new mixtures affected by higher imperatives. The more one studies business policies in the making, the more artificial do the conventional distinctions between various corporate activities appear. One might almost say that these impede a clear understanding of what top management does.

This generalisation applies most obviously to the different management functions. For example, it is inadequate to consider Dorman Long's financial convolutions under the heading of 'finance' or to analyse Stewarts and Lloyds' marketing triumphs under the heading of 'marketing'. For in practice these aspects were intimately bound up with other functions and, more importantly, with general policy. Again, studied within their respective functional compartments, the USC's productivity measures after 1928, its research and development, its planning–budgeting systems or its management training schemes would all seem curiously emasculated. Instead, viewed as strands in a combined search for efficiency, they spring vigorously to life.

The inappropriateness of functional divisions applies even more clearly to the policy-makers. No single function, not even a mixture of functions, adequately describes any of the leading managers' orientations. Thus from one angle it would be fair to claim Robert Hilton's successes as a triumph for engineering or to see Allan Macdiarmid's and Ellis Hunter's achievements as an apotheosis of accountancy. In a wider sense there can be no doubt that both engineering and accountancy played a major and perhaps insufficiently recognised part in the interwar development of management. But even in terms of these men's professional skills the role of the industries and organisations they had worked in was important, too. Hence Macdiarmid's heavy debts to Stewarts and Lloyds' corporate traditions, Hunter's to the non-iron and steel firms which he had audited, Benton Jones's to the coal industry, Hilton's to electrical engineering and Metropolitan-Vickers, not to mention professional soldiering. Moreover, all these men brought to general policy-making certain temperamental qualities and innate abilities as well as frameworks of concepts gathered from the surrounding culture. If the functional disciplines of finance, production engineering or marketing strongly influenced their methods, it was from wider sources, in the main, that they drew their guiding ideas.

A related point is that business policies were greatly concerned with values which pervaded a whole managerial régime. As we have repeatedly seen, each leading manager brought to his firm a fund of presuppositions, principles and potentials which unfolded during his tenure. Even if each major decision is correctly seen as multi- or para-functional, it is still misleading to view it as self-contained. Instead, the whole of a régime must be looked at in order to make sense of its parts.

In particular, fundamental beliefs about the purposes of business were at stake. Despite much overlap, the chief executives' emphases differed markedly here. Consider, for example, Macdiarmid's pre-occupation with the game, with conquest and winning, Hilton's with professional skills in the pursuit of multiple responsibilities, Dorman's and Benton Jones's with transcendent causes like the empire and human relations. Degrees of adherence to economic orthodoxy had some bearing, whether these were rigorous (as in the rare case of Sir Hugh Bell), half-hearted (as with someone like Sir Arthur Dorman), powerful but largely implicit (Allan Macdiarmid) or barely evident (Hilton, Benton Jones). Their attitudes towards iron and steel making as a basic manufacturing industry were germane. Most imbued it with aristocratic-cum-moralistic overtones, by the same token denigrating lighter forms of industry and, still more, the activities of the City. This could affect matters as apparently diverse as external financing, forward diversification and attitudes towards relative newcomers to the industry like Stewarts and Lloyds.

Different values could affect behaviour towards both competitors and the firm's internal organisation. For example, were one's industrial peers viewed as adversaries, to be beaten or lassoed into private alliances within a Hobbesian framework, or else as potential partners in multilateral consortia or industry-wide schemes? What views about human nature lay behind a management's relative reliance on coercion, financial rewards or the formation of *esprit de corps* as co-ordinating devices inside the firm? Albert Peech's method implied a trusting, Rousseau-like vision. Stewarts and Lloyds' emphasis on centralisation, profit-sharing and merit bonuses reflected maximising and individualistic assumptions of a Benthamite type. Hilton and Benton Jones combined central dictation with efforts to induce a co-operative atmosphere which drew both on fashionable ideas of human engineering and on older notions of organic community. Whatever the derivation of these value differences, they certainly produced deepseated contrasts in the ethos of the firms.

\*　　\*　　\*　　\*

The evidence in this book will disappoint disciples of a rationalistic model of management. It discourages the idea of a tidy, quantified, synoptic type of decision-making. However, it also lies at a marked distance from the contrary view of management as pragmatic reactors or tactical trimmers, deciding things 'sequentially', responding to successive situations merely 'incrementally'. If one takes only the documents immediately surrounding a particular decision, it would be easy – but fundamentally misleading – to slip into the latter view. Rather than generally converging towards one or other of these conceptual extremes, decision-making processes fanned out across a wide intermediate spectrum. Along that spectrum varying degrees of rationality are discernible in terms of clarity of purpose, consistency, evaluation of alternatives, research, quantification, deliberation, and the use of managerial abilities available to the firm. It is on this intermediate ground that analysis has concentrated. But the evidence suggests that managerial performance on such criteria was often closer to 'rationality' than the more sceptical 'behavioural' approaches would suggest: closer, too, than would be implied merely by the records of particular decisions or even by the topical views of managers and others, where available, on those decisions.

Thus Sir Arthur Dorman's persistent economic optimism through the 1920s, however misplaced and seemingly confused, in one sense synchronised perfectly rationally with decades of experience of relatively short economic cycles. In a way, it was highly inductive and probabilistic. Even apparently rushed and messy decisions could reflect lengthy previous consideration. For example, specific acquisition possibilities could be surveyed spasmodically over many years, it seems, long before they came to a crunch, as with Dorman Long's takeover of Samuelsons in 1917 and, in all likelihood, the SPT – Samuel Fox merger of 1916. Management could be more aware of technical advances overseas than their (often financially constrained) practices would suggest at first sight. This applies, for example, to knowledge of fuel economy in the USC in the early 1920s, of weldless-tube-making processes in Stewarts and Lloyds before 1926, of economies of scale in blast furnaces in Dorman Long. It is also easy to underestimate the extent to which decision-making took account of ranges of alternatives. Stewarts and Lloyds' successive decisions about iron and steel supplies set the pace here from at least around 1910. But dispersal in both time and documentation, even much apparent confusion, did not prevent wide ranges of alternatives being considered, as was substantially the case with the USC board's agonised heart-searchings over scrappings, divestments, reconstructions, refinancings or mergers in the early 1920s. Above all, a longitudinal perspective supports the idea that leading managers built up highly complex synoptic models of the firm and its environment. Only Allan Macdiarmid and Walter Benton Jones got close to some comprehensive written formulations of these models. For the other men they emerge in bits and pieces and were largely informal or implicit, but for all that hardly less substantial.

If a long historical view throws strong doubts on the notion of managers as 'incrementalists', tacticians or parochial pragmatists, it also severely qualifies our definitions of managerial failure or success. In the case of each major decision a long backward view is needed to understand its provenance and nature; but a long forward view is essential, too, in order to observe how well or badly it turned out. This is partly because defects and changing fashions in public opinion could obscure key aspects of performance. For example, even a sophisticated observer like Montagu Norman, voicing the current cult of giant mergers, saw the USC in 1930 as falling short on 'rationalisation' precisely when it was in the throes of an outstanding internal reformation. Neither the financial press nor concerned City interests came near to understanding Dorman Long's financial quagmire until the early 1930s, a tribute to the frequent superficiality even of expert current opinion. But a long time perspective is even more necessary because of the effects of economic fluctuations, particularly in a capital intensive, economically sensitive industry like iron and steel (and, it should be added, in other large-scale industries which have developed since).

Such a perspective obviously qualifies the very definition of 'error'. For example, Dorman Long's Redcar project, the USC's Templeborough complex and that firm's expanded commitments to plate-making at Appleby, were all products of the 1916–20 economic optimism which turned badly sour during the Depression. But they proved reasonably successful by the mid-to-late 1930s as economic conditions revived. Even the USC's West Cumberland imbroglio, although stemming from serious misjudgements, produced marked social benefits and, from 1928 onwards, as the firm shouldered the burden more explicitly and intelligently, it became a managerial achievement of considerable merit.

The key concept, as we have repeatedly seen, is that of avoidability. Did the decisions turn out badly within the usual time scale of planning? Could management have reasonably foreseen that they would? The former criterion means excluding, for instance, the failure of the Stewarts and Lloyds' board in the 1930s to envisage that its lack of forward diversification would lead to problems over both product competition and nationalisation by the 1950s and 1960s. The latter test means taking careful account of the information available at the time the original decisions were made. The problems of evaluation as well as measurement should not be underestimated. Nonetheless a broad typology of the avoidable errors covered in this study does emerge. Of the following categories (1) is the least culpable whilst (4) remains the hardest to evaluate:

1   *Participation in collective errors*: large or small misjudgements where management mainly reflected general current opinion about macroeconomic prospects (the massive expansions of basic steel-making capacity between 1916 and 1918).
2   *Minor own-sector errors*: misjudgements within management's areas of knowledge and competence involving costs which were small relative to the firm's resources (Stewarts and Lloyds' preference for the NLIC in 1917, its tactical errors over the Corby project, and Dorman Long's mishandling of aspects of the South Durham negotiations, 1930–3).
3   *Major own-sector errors of commission*: misjudgements within management's field which involved relatively large commitments of resources (the new USC's investment in West Cumberland in 1917–18, Dorman Long's £3½ million debenture issue in 1923, Sir Arthur Dorman's failure to retire after 1918 which substantially committed the firm's fortunes to a single old man).
4   *Major own-sector errors of omission*; failures to take advantage of major opportunities, again within areas proper to management (the USC's failures on strategic expansion in the Midlands in the 1930s, and the organisational efficiency shortfalls in the USC up to 1928, in Dorman Long up to the mid-1930s).

Of these errors perhaps the most outstanding was Sir Arthur Dorman's failure to retire. Dorman's remarkable tenacity and partial achievement even during this period do not outweigh the tragic character of his persistence. It can be argued, of course, that the lack of suitable succession material left him with no alternative. But this very gap reflected a failure to renew the board, to delegate, to develop new talent or recruit new blood. Dorman Long's 1920s syndrome of elderly dominance, proud traditionalism, absentee grandee directors, primarily dynastic promotions at the top and general managerial inbreeding, exemplifies the cumulative troubles which can afflict family firms at certain stages. At another level it provides a mordant example of the managerial component of what has been called the 'English disease'. It is only within this wider context that the firm's more specific errors over financial, organisational and merger matters can be fairly understood, not least the South Durham merger breakdown in 1933.

\* \* \* \*

Managerial concentrations on growth and efficiency, and on the trade-offs between these two objectives, have absorbed much of our attention (leaving aside social action for the moment). One part of this theme is familiar: the frequent tendency for attempted combinations between the two to come unstuck for reasons outside management's immediate control. We have repeatedly seen how managements genuinely sought to unite growth and efficiency, particularly in terms of production facilities and through the agency of both investment projects and mergers. We have repeatedly observed, though, how the two desiderata became involuntarily split. Into this hard frictional category fell not only the distorting effects of recessions on capital investment projects but also some less often noticed phenomena, notably the confusions and duplications engendered by rapid growth through merger, and the time conflicts, often exacerbated by geography, which made it hard for management to pursue growth and efficiency simultaneously in properly intensive ways.

However, it is the idea of managerial specialisation as a frequent source of bias, an argument more distinctive to this study, which invites the most debate. For at first sight this appears doubtful or at least a factor which should be softer and more malleable. Inevitably, questions arise as to how fundamental, deep-rooted and typical such specialisations were, and whether they could have been avoided or offset.

The objections to viewing growth and efficiency as primary and often rivalling managerial pursuits arise from several quarters. There is the profit maximising theory which would subjugate both to its own all-enveloping monism. But we have repeatedly seen that, despite its central importance as a constraint, a yardstick and an instrument, profit was not the supreme force intellectually, emotionally or morally. There is the claim that growth overshadows efficiency or even everything else (another type of monism) or that it shares the throne co-equally with altogether different goals. But we have repeatedly found that managerial ideas about efficiency in a wide sense, going well beyond just production efficiency, had a life of their own. Still more so did certain powerful inclinations along the same path. Again, there is the pluralistic objection that managerial differences are so great that they cannot properly be subsumed within two or three central trends. But while no theory can be exhaustive, the undoubtedly rich differences of the leading managers examined in this study do appear to have crystallised to a considerable extent around propensities for growth and/or efficiency. Finally, there is the denial that, apart from the enforced conflicts already mentioned, managerial propensities significantly or frequently polarise on one to the serious detriment of the other. It may be said that, in general, managers sincerely seek to combine growth and efficiency (as if desires rather than capacities were all-important); that they can learn over time to make up for deficiencies in one or the other (as if managerial capacities were largely malleable); or that one man's limitations here can readily be compensated by another's strengths (as if managerial equalities, synergies and power sharings were typical, and also as if underlying value conflicts did not arise).

There can be little doubt that individual managers' underlying motives first of all for growth were remarkably deep-seated and consistent. Below the familiar surface features of economic optimism and enthusiasms for technical progress lurked more elemental phenomena: expansionary environmental influences during earlier formative years, a liking for power and money, a gambling streak, a zest for the extrovert side of business, chauvinist and imperialist feelings or, especially in Allan Macdiarmid's case, the idea of business conquest as a game and the sheer will to win. It was factors like these which ensured that the drive towards growth was no transient plant, obeying the changing economic weather, but a persistent behavioural tendency.

The strength of personal thrusts towards efficiency was partly a response to another factor much emphasised in this book, the persistence of the efficiency problems of the large enterprise, even in a sense their unresolvability. A great amount could be achieved relatively quickly, as in the USC between 1928 and 1932, but it required an enormous continuing effort to sustain. The problem of slack was endemic. As a Stewarts and Lloyds official put it in 1936, even as massive efficiency gains from Corby were accumulating: 'The hunt for lost time is always on.' More fundamentally, the problems of centralisation versus decentralisation admitted of no facile answer. In retrospect it is easy to argue that the USC before 1928 was too decentralised, that Dorman Long before 1931 was too centralised in some ways, too decentralised in others or that Stewarts and Lloyds' ultra-centralisation by the 1930s was unsustainable and had some major costs as well as benefits. It is tempting to smile at the attendant conceptual and semantic confusions, although on these issues it should be admitted that we have not advanced much since then. It is harder by far to see how an optimal equilibrium could have been defined, let alone attained. Perhaps the best norm would have been a continuous experimentation in the avoidance of extremes which is, in fact, roughly what the USC did in the 1930s, albeit jerkily and unclearly.

However, for purposes of the argument about managerial specialisation, the important point is that the efficiency imperatives of large-scale organisation demanded responses of a deep-seated and personally abiding kind. Here the familiar economising instinct, the will to maximise, although not purely in the would-be *profit* maximising sense, occupies much of the foreground. But parallel with or behind this factor, even in a sense absorbing it, were concepts particularly relating to tidiness or symmetry, discipline or precision, command over men (including a component of sheer 'bossiness'), ideals of harmonisation and human development. The marks of the engineer and the accountant were evident here. So were the lurking propensities of the aesthete, the orchestral conductor, the dictator, the doctor, the teacher. As we have seen, the search for such attributes in the case of the main efficiency proponents in our story, notably Hilton, Benton Jones and Hunter, yields rich rewards.

It is hard to see how these contrasting propensities for growth and efficiency could have changed. Sir Arthur Dorman's growth-manship endured through the long depression years. Harry Steel's might well have done had he survived. Allan Macdiarmid's dedication to both pursuits was coterminous with his tenure. Hilton and Benton Jones's biases towards efficiency and against strategic growth did not disappear even when economic recovery made the former appear less urgent, the latter more feasible. Well into middle age or beyond, constant in their values, set in their ways, justified by their previous successes, these men could not alter their basic orientations at will.

Of course, even if the existence of such biases is accepted, it may still be argued that they should have been corrected by the construction of better-balanced management teams. But this makes unrealistic assumptions about the ability of strong individuals to recognise their own limitations, to recruit speedily in the required direction and, above all, to share power co-equally and harmoniously when it was a question of differing personalities, varying propensities, even divergent ideals. The integration of Stewarts and Lloyds' inner circle of the 1930s was surely exceptional. It relied on a rare combination of ethnic–cultural affinities, family bonds, friendships and contiguities. It was more like a royal court than a 'team'. Overall, the extent of concentration of power has been repeatedly observed in this study.

Given the various constraints, it is hard to see how Sir Arthur Dorman could have brooked a near equal master-minding an efficiency drive in the 1920s or how Charles Mitchell could have tolerated, or for that matter imposed, such a man's necessary dominance over Middlesbrough in the early 1930s. It is no easier to envisage the USC's dual (efficiency-orientated) monarchy from 1928 transforming itself into a triple (efficiency- plus growth-orientated) one. For some time Hilton and Benton Jones's urgent priorities in working together and getting on top of the organisation would have been diluted, to say the least, by the entry of a powerful third man with different interests and characteristics – even assuming one could be found and, if recruited from outside, quickly trained. Later, such a course should have been easier in theory as well as more desirable, but by then the new habits had become ingrained. Hilton was unlikely to cede power to a co-equal; the advent of a new top man from the outside would have been resented by influential up-and-comers in the second layer; and a style of aggressive growthmanship, additionally implying separatism on federation–public issues, would have subverted the USC's corporate ethos.

It may still be objected that these were special cases, that other outstanding managers have combined strong propensities towards both growth and efficiency, and that the underlying drives are not typically irreconcilable. Certainly, the specialisation theory is not proved by a few closely worked examples like these. Nonetheless the evidence is suggestive and the hypothesis of typical managerial polarisations towards growth or efficiency still seems plausible. So does the related hypothesis that these were likely to underlie similar phases of corporate behaviour in so far as power was concentrated.

*     *     *     *

The third main contributor to policy trade-offs and biases was not co-equal with growth and efficiency. But social action could, none the less, have important effects on a firm's behaviour, and also on the social environment. In the latter sense even the local impacts of, say, Dorman Long's civic links and philanthropy in the 1920s cannot be ignored. More strikingly, the USC's social action pulled it to the forefront in advocating industry-wide reforms between 1928 and 1935, helped to save tens of thousands of people from poverty and insecurity by preserving a whole area from economic devastation (the West Cumberland case), assisted in bringing joint consultation and retirement pensions to many thousands of workers, and notably contributed to the industry's emergent 'Middle Way' in social control. Conversely, shortfalls in social action, notably Stewarts and Lloyds', adversely affected the physical environment, redundant workers, and central efforts at planning.

For various reasons the underlying drives tended to be understated in the records. But as with growth and efficiency, they were persistent. Local–civic bonds; quests for social recognition; super-charges of chauvinism; concepts of a larger 'good of the industry'; the ideologies of class harmony, job fulfilment and wealth-as-stewardship (evident in Benton Jones's case); an actively searching, not merely constraint-accepting sense of the growing importance of government, consumers and public opinion (a key factor for Hilton): such propensities spanned the managerial tenure. So did contrasted blind spots to anything much beyond growth and efficiency sectionally-conceived.

Sometimes, of course, social action did blend happily with pursuits of growth or efficiency. When Sir Arthur Dorman and Harry Steel rushed to exploit domestic raw materials in Kent and Cumberland, war-reinforced feelings about national self-sufficiency lent an extra thrust to their commercial growthmanship. When the USC and Stewarts and Lloyds increased current welfare spending, a humanitarianism agreeable both to their own consciences and to public opinion seemed to march hand in hand with enhanced labour productivity. When all three firms gave money to charities, more or less generously, no real conflict with profitability existed since the sums were so small, although in the 'stingy' case of Stewarts and Lloyds in the 1930s such conflict may have been felt.

However, the growth and efficiency effects of bigger forms of social action could be speculative, even adverse, and be anticipated as such. For instance, any cash benefits from the USC's joint consultation and pension schemes rested on a hypothetical causal chain through morale to productivity that was debatable. The firm could not expect much, if any, special commercial return from its vigorous espousal of industry-wide collective interests and public responsibilities. The benefits here were anticipated as partly political, a safeguard for continued tariffs and semi-independence, and partly, therefore, economic in the longer term, but in both cases collective. Again, the USC's maintenance of West Cumberland, strongly influenced by considerations of duty and social repute, patently reduced profitability and opportunities for growth in other fields. More subtle were the possible implications of a state of mind that was anti-rapacious, temperate on competition, basically non-aggressive. In the USC of Hilton and Benton Jones such attitudes were perfectly compatible with a consistent, indeed distinguished pursuit of efficiency, and also with moderate growth. But they accorded ill with a profit-maximising species of efficiency which would, for example, have created massive redundancies, and with a pushful, displacement-creating brand of growthmanship.

The conflicts are equally highlighted by the record of Stewarts and Lloyds in the 1930s. Given the various opportunities and constraints, it is hard to conceive of a better performance on growth and efficiency (although in the latter case extreme centralisation raised some question marks about the longer term). But on social action the firm was deficient. Would a better record here have stymied its economic brilliance? It is hard to say. On the one hand, perhaps the firm could have opted for co-operation in the development of Northamptonshire iron ores without losing its competitive edge in the tube industry. It could probably have afforded rather more philanthropy, redundancy compensation, environmental care and even domestic price restraint, leading to somewhat lower profits than the bumper ones it actually made. On the other hand, there was probably some real incompatibility over underlying drives and states of mind. In the last resort, it seems doubtful whether the requisite social vision and moderation were reconcilable with the single-minded forcefulness which made Allan Macdiarmid so formidable.

Whether such conflicts were typical is harder to evaluate. In all likelihood, the characteristics which most favoured social action were inconsistent with a quantophrenic attitude towards gain which disparaged intangible pay-offs. They conflicted with a highly-positive time preference which demanded immediate returns. They probably clashed with an obsession with the firm's relative position which despised advances shared proportionately or equally with others. Nor do they seem compatible with 'free rider' attitudes, extreme individualism and a form of sectionalism that bent to public–political considerations only when these were enforced or the subject of immediate and palpable threat.

The wider implications of such conflicts for both the steel industry and management generally deserve some brief comment. On the one hand, there is the controversial question of what is, or would have been, the best organisation for iron and steel. An industry split up and run on rigorously inter-firm competitive lines? Unitary nationalisation? Some kind of middle way? The answer will strongly influence views on how far managerial specialisation and corporate performance biases as between social action and the other pursuits should be taken seriously.

If rigorous competition is regarded as viable and preferable, much social action of the USC type can only be lamented, in unison with the Chicago School, as blunting the required sharpness of sectional profit-seeking. If unitary nationalisation is espoused, voluntary co-operation with industrial and public policies by sectional interests ceases to be important, at least in theory (although it hardly vanishes completely, given the impossibility of absolute central control). However, where compromises are sought, social action becomes indispensable. For example, a publicly supervised oligopoly or a mixed ownership solution may be pursued as the best way of avoiding the pitfalls of uneconomic competition, cartel exploitation, over-centralisation and excessive politicisation. But in such a middle way there is a strong need for the sectional units to co-operate with a decentralised form of social control.

The dilemmas posed by the contrasts between Stewarts and Lloyds and the USC in the 1930s have implications well beyond the steel industry. They raise questions about society's expectations of business in general. There are strong grounds for saying that in the 1930s government policies and public opinion not only expected business to be efficient and dynamic and, in the case of the basic industries, to participate in collective reorganisation and planning schemes. Also, even outside the basic industries, the cause of bigness was exhorted in the interests of 'rationalisation'. So were non-exploitation and commercial restraint. So, for that matter, were conciliation of labour, sensitivity to unemployment, economic nationalism and environmental protection. The conflicts between these various desiderata were greatly underestimated by people outside industry, from Whitehall downwards. But this failure to reckon with the difficulties faced by businesses in reconciling different public interests has hardly receded since.

Of course, some of the interwar calls on firms have receded. In particular, the exhortation to protect jobs and help redundant workers became less marked after 1945 in so far as full employment became the state's task and collective provisions for the unemployed improved. Conversely, though, new priorities for social action emerged, for instance with regard to pollution, restraint on pay increases and various forms of business–government co-operation in a mixed economy weakened by decline and threatened by crisis. Moreover, one basic rationale for such action – the unfeasibility and undesirability of reducing all social desiderata to legislation or government *diktat* – appears likely to persist. The social sensitivity and public co-operation required from private enterprise continue to imply that extreme sectionalism on efficiency or growth will be disapproved. But the theory of managerial specialisation, as confronted in this study, including the tensions uncovered in the three firms analysed, strongly suggests that the value dilemmas should be recognised more fairly and honestly.

*     *     *     *

What does this study suggest about the long-term development paths of large firms? The management phases our three firms went through, as interpreted in the book, are summarised below. The sequences of these phases should be noted: in the first two firms growth–efficiency alternations (with intermediate periods of indeterminacy), in the last a growth plus efficiency reiteration. How far did these sequences reflect economic fluctuations or internal corporate–managerial processes? What wider lessons, if any, can be drawn from them?

Table 10.1   *Corporate development phases 1914–39*

|  | Period | Chief executive | Phase category |
|---|---|---|---|
| *Dorman* | 1914–31 | Sir Arthur Dorman | G |
| *Long* | 1931–34 | Charles Mitchell | I |
|  | 1934–39 | Ellis Hunter *et al.* | E |
| *USC* | 1914–20 | Harry Steel | G |
|  | 1920–27 | Albert Peech | I |
|  | 1928–39 | Robert Hilton and Walter Benton Jones | E and S |
| *Stewarts* | 1914–25 | J. G. Stewart | G and E |
| *and Lloyds* | 1925–39 | Allan Macdiarmid | G and E |

The obvious point that the economic situation strongly influenced the degree of success of these phases has already been conceded. In favour of the economic responsiveness argument it can also be claimed that the increasing emphasis on efficiency-seeking through the period reflected the deep shocks administered by the Depression. Nonetheless there is a strong case for seeing the sequences as fundamentally corporate–managerial phenomena.

If the phase changes really represented managerial adjustments to new economic conditions, the time-lags appear surprisingly acute. Despite the severity of the Depression it took Dorman Long at least fourteen years to switch unambiguously to an efficiency phase, the USC about eight years. The feedback mechanism, if such it was, was unconscionably slow. Moreover, most of the phase changes were not strictly in tune with current or prospective economic conditions. By the time of Dorman Long's switch in 1935–6, economic prosperity was clearly returning so supple responsiveness would have implied, if anything, renewed growthmanship. Stewarts and Lloyds in 1925–6 did not strictly obey economic conditions by shifting to an efficiency concentration. Instead, growth as well as efficiency moved upwards on the managerial agenda. Only the USC's change to an efficiency concentration in 1928 seems to have been reasonably cycle-sympathetic, albeit belated. Even then it was qualified by the factor of a new emphasis on social action.

Even if the influence of economic fluctuations on the sequences was weak, as the evidence suggests, this does not in itself prove that internal corporate–managerial laws were at work instead. A major difficulty in observing any such laws is the occurrence of periods of indeterminacy whose timing was probably random. On the other hand, as we have seen, all three firms' sequences can reasonably be interpreted as following an internal logic.

First, Stewarts and Lloyds' success in one growth plus efficiency phase had a clear result when the managerial régime had to be changed in 1925–6: the instalment of a new régime promising to do the same thing better. Of course, that promise could have been unfulfilled but at least it showed a strong corporate desire to stick to a growth-efficiency combination once this had been reasonably approximated. Second, both Dorman Long's and the USC's growth concentrations led to confusion and imbalance which cried out for redress through an efficiency phase. This was clearly perceived by some leading managers and, in Dorman Long's case, also by influential outsiders; and the consequent reaction played a major part in bringing about the changes of 1927 in the USC and of 1931 and 1935–6 in Dorman Long. A pivotal point is that in all likelihood such a reaction towards efficiency would have occurred even if the economic situation had been more benign. Third, no case of a contrary swing from efficiency to growth was observed during the period. But it seems reasonable to assume that an efficiency concentration would eventually produce its own corporate imbalances and internal reactions; and that endogenous pressures towards a growth phase would emerge, although once again not effectively so until the conditions for a managerial changeover were ripe (death, natural retirement or a boardroom coup).

Overall, therefore, our case studies are consistent with, indeed lend some support to, the two underlying hypotheses about long-term development paths: a straight-line reiteration and a zig-zag alternation, both primarily reflecting corporate–managerial forces.

Whether such patterns have been typical before 1939 or more recently is another matter, as is the question of their relative frequency. Alfred D. Chandler's observation of a frequent growth →consolidation→growth→consolidation sequence pertained to the formative period of large firms in the USC in the late nineteenth and early twentieth centuries, although it is still highly suggestive. The central problem is, of course, the historio-graphical difficulty of putting large businesses under the micro-scope, particularly for more recent periods. Some observable trends at least have an indirect bearing, for example the increas-ing size of firms and the substitution of professional management for the family influence in large businesses. Both have probably led to a reduction in the length of top management tenures.

It is tempting to assume, optimistically, that a collective man-agerial learning process has occurred; that management training and development have produced a greater supply of excellent all-rounders; that the art of creating balanced management teams has greatly improved; and that large firms have moved towards smoother development patterns. Such a progressivist, Whiggish theory is attractive to some management theorists. It also appeals, no doubt, to those ideological defenders and critics of the large modern corporation who have a common interest in seeing it as managerially successful.

But the idea that managerial conflicts are progressively on the wane has no empirical support. There is no reason to believe that phases of indeterminacy occur less often. Indeed, there are grounds for suspecting that the understandable factors of corporate secrecy and sensitivity, coupled with the inadequacy of proper evidence, lead to their frequency being underestimated. The hypothesis of corporate concentrations on growth or efficiency cries out for systematic verification. Meanwhile, the tendencies suggested as favouring that hypothesis appear very plausible – namely per-sonal behavioural biases among top managers, concentrations of power on single individuals or tiny groups, and acute time press-ures on the latter. So do the arguments for envisaging long-term reiterations $(G + E \rightarrow G + E \rightarrow G + E...)$ or alternations $(G \rightarrow E \rightarrow G \rightarrow E \rightarrow ...)$. The tendencies for corporate success to be self-reinforcing to some degree (supporting the former) and for strong thrusts in one direction to lead to dialectical corporate reactions (supporting the latter) are in all likelihood deep-seated.

The more radical idea that zig-zag development paths may actually dominate over straight lines also remains to be verified. Within any comparable group of large firms, substantial differences in growth rates over fair periods of time, say ten years, are highly likely. Investigation may well show that above-average growth reflected an intense policy concentration in that area and that below-average growth reflected an absorption with efficiency measures. It is also likely that the firms which lead or lag in one 'long period' are different from those which lead or lag in another, providing *prima facie* evidence of inter-temporal shifts of behaviour in the same firms over the 'very long-run'. This, too, should be checked, although the difficulties would be greater. It is hard to see how the most remarked-on trends affecting modern business organisation, those towards increasing size and diversification, would undermine such tendencies. On the contrary, increasing size may well accentuate efficiency problems. For its part, increasing diversification doubtless evens out short-term fluctuations in growth and profitability, as intended. But diversification is unlikely to destroy either inter-firm differences in long-term growth rates or the problems of organisational size and complexity, repeatedly identified in this study, which lead to substantial pauses for breath or absorption. On the contrary, again, diversification may even exacerbate such problems. Nor does it seem likely that improving management techniques, for all their refinement and apparent sophistication, have massaged out such differential spasms and discontinuities.

*   *   *   *

If a central ideal about management emerges from all this it is that of balancing a few profoundly creative, socially approved pursuits in the long run. Ideally, simultaneous excellence in growth, efficiency and social action would be attained. But for a variety of reasons, some doubtless good (the tendency of human abilities to specialise), others at least ambivalent (time constraints, enforced splits between the pursuits, power concentrations), such an outcome must be accepted as rare. More often an attainable ideal would be alternating phases of excellence in either growth or efficiency but with each accompanied by moderate success in the other and by continuing social action.

The alternation would mainly reflect a key assumption in the above argument, that of frequent biases in the creativity of the sort of outstanding individuals who are badly needed to make industry dynamic. The norm of continuous social action would recognise management's part in building that element of free co-operation and solidarity without which either social disunity or state dictation would rule. But none of the three pursuits would ever be lost sight of. Management's task would be comparable with reconciling conflicts of vital aims in a whole modern economy, not with a single obsession. Monism would be banished both as an interpretation of what management does and as a precept. Whatever its theoretical uses in economics and its attractions to seekers after tight, simple certainties, the idea of single objective maximising has always been a diversion from normative model-building around the real-life tensions and value aspects of management.

The idea of balancing several objectives has some unfortunate connotations of looseness or excessive pragmatism. But this is not necessarily justified. So much depends on what are the prescribed objectives. If the specification is to reconcile conflicting interests or pressures, for example those of employees, shareholders, government, etc., this could imply mere tactical *arbitrage* and politics, trimming and satisficing, 'rubbing along'. If the objectives are vague and unquantifiable, if standards of performance cannot be set, management will be let off too lightly. Moreover, if the balancing is seen as purely technical, a matter of functions and formulas, the unifying grandeur of business policies is lost, management is belittled, the value aspects of its choices are buried. But this book has surely shown that the pursuits of growth, efficiency and social action were not like any of this. On the contrary, they were often demanding and rigorous as well as magnetic; and balancing them in difficult situations called for considerable moral courage.

# Source Material

## A  Unpublished sources

Major sources in this category have been starred and given an abbreviation, e.g.* Dorman Long, Directors' Minutes, D/DM.

### 1  DORMAN LONG

All records are at the British Steel Corporation Northern Regional Records Centre, Middlesbrough, unless otherwise stated.

|  |  |
|---|---|
| * Directors' Minutes (BSC(UK)/SEC/3) | D/DM |
| * Annual General Meetings, proceedings (1066/24/2) | D/AGM |
| * AGMs, chairman's speeches verbatim, 1892–1914 (1066/24/2) | D/ACS |
| * Financial files, 1922–30 (1066/9/1–7) | D/FF |
| * Finance Committee Minutes, 1930–4 (1066/13/1) | D/FC |
| * Private Minute Book 1, 1931–8 (1066/13/3) | D/PM |
| * Colliery Committee, 1923–9 (1066/21/18) | D/CC |
| * Correspondence relating to investments (1066/5/1–16) | D/CI |
| * Correspondence, etc., on South Durham merger (234/1/33) | D/CM |

Miscellaneous Correspondence (1066/8/16)
Data on directors (1066/8/17) (DTR 1 and DTR 327)
Chief Chemist's letter books (203/1/77)
Meetings of various groups of shareholders (1066/16/46)
Publicity Department, 1935–6 (1066/15/11)
Office Administration Committee, Minute Book (1066/13/2)
Britannia Works (803/1/2)
Sydney Harbour Bridge Contract (1066/8/3)
Bridge Department (1066/8/4)
Managing Director's Monthly Reports, 1936–8 (1066/16/11)
Papers on formation of Dormanstown Estate (1066/21/4)
Papers on amalgamation with Bell Brothers, 1923–9 (1066/25/6)
Pollution of River Tees (212/1/2)
River Pollution (210(b)/3/15)
North East Coast Steelmakers, 1926–39 (260/1/1/2/13/1)
North East Coast Steelmakers, Minutes (210(d)/15/22)
Bell Brothers, Reports, Accounts, etc. (1066/24/1)
North Eastern Steel Company, Minute Book 5, 1914–35 (1066/13/7)
Carlton Iron Company, Minute Book (1003/19/5)
Sir B. Samuelson and Company, Minute Book (1066/13/7)
P. H. Andrews, Report on Tees-side Blast Furnaces 1928 (by courtesy of Dr J. K. Almond).

### 2  STEWARTS AND LLOYDS

All records are at the British Steel Corporation East Midlands Regional Records Centre, Irthlingborough, Northants, except those listed as 'Glasgow' which are at the Scottish Regional Records Centre, Tollcross Works, Tollcross Road, Glasgow.

* Board Minutes, vols 7–12, 1914–39 (065/1/3)          S/BM
* General Purposes Committee (1791/1/14)          S/GPC
* Managing Committee (1791/1/14)          S/MC
* Office Committee (1791/1/15)          S/OC
* Works Committee (1791/1/17)          S/WC
* Excerpts from Committee Meetings for Works Managers
  (Glasgow, 666/4/1–3)          S/EC
* Confidential booklets (065/1/6)          S/CB

Miscellaneous material on committees (1791/1/14–17)
Stores Buying Committee (002/6/1–3)
Corby Labour Committee (127/1/1)
Subscriptions and donations (002/2/25)
Miscellaneous material and cartels (1791/1/21)
Notes on recent developments for staff, 1924 (1791/1/23)
Report to Lord Weir on manufacture of iron and steel, May, 1929
(065/4/3)
Economy Commission (1791/1/21)
Corby Costs, 1934–9 (1687/1/1)
Lancashire Steel Corporation (002/2/1 and 4)
Stanton Merger, correspondence file (Glasgow 1207/18/11–12)
Stewarts and Lloyds (Australia) Ltd (1791/1/33–4)
Stewarts and Lloyds (South Africa) Ltd (1791/1/35)
Scottish Tube Company, AGMs, 1919–31 (1791/6/16)
Scottish Tube Company, Private Ledger 1 (002/2/13)
John Spencer Ltd (065/2/3)
Islip Iron Company (002/1/10)
Kettering Iron and Coal Company (065/3/4)
Bromford Tube Company (065/2/4)
Scottish Tube Makers' Wages Association (Glasgow 680/1/1–2)
Chairman's Conference, May 1962, Talks, Stewarts and Lloyds Ltd (by
courtesy of Mr E. G. Saunders)

## 3  UNITED STEEL COMPANIES

All records are now at the British Steel Corporation Northern Regional Records Centre, Middlesbrough, unless otherwise stated.

* Directors' Minutes, 1918–39 (BSC(UK)/SEC/2/1–4)          U/DM
* Central Committee Minutes, vols 1–14, 1920–39 (159/3/3,
  7 and 8)          U/CC
* Finance Committee Minutes, 1920–8 (159/5/1)          U/FC
* Finance Committee, correspondence and papers, 1921–5
  (159/5/1)          U/FC/C
* Annual General Meetings (159/1/1)          U/AGM
* Workington Works Directing Committee, 1920–7 (159/3/1)          U/WWD
* Sir Frederick Jones's correspondence, 1920–30 (006/24/37)          U/FJ
  * Reorganisation of administration and Mr Hilton's
    agreement (006/24/37)          U/RA

\* Iron ore: consolidation of resources of United Steel and other
  companies, 1923–37 (159/5/2)                                    U/IO
\* Chairman's investigations (Sir W. Benton Jones) (152/11/21)    U/CI
\* Copy letter books from Managing Director (R. Hilton),
  4 books, 1929–38 (167/41/3)                                     U/MD

Amalgamation with various concerns, 1921–3 (006/10/3)
General Matter file, 1924–9 (006/10/3)
Amalgamation 1927 (006/10/3)
Report on the properties and operations of the USC by H. A. Brassert
and Company, September 1929 (006/10/1)
Chairman's Finance Policy Memorandum 1933 (006/10/1)
Notes on investment and profits (006/10/1)
Long-date budgets (152/11/4)
Office Manager's Minutes, 1935–41 (152/11/2)
Manufacturing costs, SPT, S. Fox, Appleby, etc. (152/11/12)
United Steel Companies Ltd, Report by Arthur McKee and Company,
1939 (on the Workington Branch) (by courtesy of Mr J. Lancaster)
Summary history of the Rother Vale Collieries, 1862–1950 (159/5/2)
Steel, Peech and Tozer, General Minutes, 1908–31 (159/3/3)
Workington Iron and Steel Company, Directors' Minutes, 1909–18
(Glasgow unlisted)
Net profits from 1870 (006/10/1)
The history of Appleby Frodingham, an outline by G. R. Walshaw and
C. A. J. Behrendt, 1950 (159/5/1)

## 4 OTHER COMPANIES

*Bolckow Vaughan (BSC Middlesbrough)*
  Directors' Special Minutes, 1921–9 (1066/13/7)
  Historical material (210(d)/20/50)
  Upton Colliery (260/1/1/4/15)
  Report by Sir W. Peat and Sir W. Plender, August 1924 (1066/25/5)

*Colvilles (BSC Glasgow)*
  David Colville and Sons, Minute Books 5–6 (1283/1/9–10)
  National reorganisation, 1931–4 (1283/1/5)
  Mr Craig–Stewarts and Lloyds (1283/1/7)

*Consett (BSC Middlesbrough)*
  Minute Book 24, 1934–7 (CHM/SEC/1/2)
  Reorganisation of iron and steel industry (218/6/33)

*South Durham and Cargo Fleet (BSC Middlesbrough)*
  Directors' Minutes, 1914–39 Vol 1 (1066/13/1)
  Vols 2–4 (208/1/1/1)                                            SD/DM

  Additional materials on merger, 1933 (234/1/33)
  Papers of W. A. Caddick PD/1
  Cargo Fleet (210(d)/15/18)
  Office Manager's files (210(d)/19/16)

*Richard Thomas and Company (BSC Newport)*
  Minute Books 2–5, 1914–39 (271/3/31 and 32)


## 5   FEDERATION RECORDS

All records are at the British Steel Corporation East Midlands Regional Records Centre, Irthlingborough, Northants, except those listed as 'London' which are at the BSC Headquarters. Data refer to the National Federation of Iron and Steel Manufacturers (NFISM) up to 1935, the British Iron and Steel Federation (BISF) from 1935 to 1939.

* Executive Committee Minutes, 1918–35 (802/6/11) ⎫ F/EC
* Executive Committee Minutes, 1935–39 (802/6/12) ⎭
* General Meetings, 1919–39 (802/6/4)                         F/GM
  Verbatim reports of AGMs, 1919–35 (802/6/59)        F/AGM
  Parliamentary and General Purposes Committee Minutes, 1918–39 (802/6/5)
  Film 3 (includes lobbying for tariffs, 1931–2) (1790/8/1)
  Film 4 (includes data on prices, investment and tariffs, 1935–9) (1790/8/1)
  Film 39 (includes data on National Committee, 1933) (1790/8/1)
  Films 12, 15, 19, 28, 37 (1790/8/1)
  Iron, Steel and Allied Trades Federation, Statistical Reports for 1915, 1916 and 1917 (London, Head Office Library)


## 6   BANK OF ENGLAND RECORDS

All records are in the Central Archives Section of the Bank of England, Threadneedle Street, London. Abbreviations: Securities and Management Trust, SMT; Bankers' Industrial Development Company, BID; Bank of England, BE. 'Chairman' refers to Montagu Norman.

Chairman's Papers Iron and Steel, 1927–34 (SMT 2/72)
United Steel and John Summers, 1938 (SMT 2/63)
Chairman's Papers, Dorman Long and Company (SMT 2/93)
United Steel Companies, 1929–46 (SMT 2/147)
Stewarts and Lloyds, 1930–45 (SMT 2/148)
Sir Walter Benton Jones, amalgamation scheme, 1936–7 (SMT 2/149)
Rationalisation in iron and steel industry, 1930–44 (SMT 2/159)

C. Bruce Gardner, notes of interviews, 1930–2 (SMT 3/18)
Iron, Steel and General Engineering, 'General' file (SMT 3/73)
Corby progress reports, 1933–5 (SMT 3/230)
IDAC and Iron and Steel Committee (SMT 3/89)
United Steel Company's Dispute with Stewarts and Lloyds (SMT 3/244)

Material on South Durham merger scheme (SMT 6/4)

Midlands, rationalisation of steel, 1930–1 (SMT 7/1)
Memorandum on heavy steel industry by R. S. Hilton, 1932 (SMT 7/2)

Meetings – Weekly (Industry), 1930–8 (SMT 9)

Proposed coke oven plant at Cleveland (BID 1/79–81)
Merger with South Durham Iron and Steel Company, 1931–2 (BID 1/82–4)

Financial Negotiations with Stewarts and Lloyds, 1930–3 (BID 1/100–106)
United Steel Companies, proposed debenture issue (BID 1/116)
Report on the structure of the iron and steel industry of Great Britain, incorporating plans for rationalisation, Charles Bruce Gardner, December 1930.

## 7 PUBLIC RECORD OFFICE PAPERS

Of the following records those relating to the Import Duties Advisory Committee (IDAC), are subject to a 'fifty-year rule' and therefore require special permission to consult.

*Import Duties Advisory Committee*
Minutes of Committee Meetings, 1932–9 (BT/10)
Committee Papers, 1932–9, 4 vols for each year, papers on iron and steel (BT/10)
Miscellaneous papers (BT/10)

*Iron and Steel Industries Committee* evidence, 1916–18 (BT/55/38)
Reconstruction of industry, financial facilities (BT/56/14)
Iron and steel trade, general position (1930) (BT/56/21)
Rationalisation of industry, general memoranda (BT/56/37)
Iron and steel trade – proposed establishment of new Coke Oven Plant for Dorman Long amalgamation (BT/5637)
Cabinet papers
Cabinet conclusions

*Economic Advisory Council, Iron and Steel Committee* (Sankey)
Report, May 1930 (CAB/CP 189/30)

## B  Published sources: books and articles

C. B. J. Maclaren, Lord Aberconway, *The Basic Industries of Great Britain* (London: Ernest Benn, 1927).
G. C. Allen, *British Industries and Their Organisation* (London: Longmans, 1970).
J. K. Almond, 'Iron and Steel' in G. Roderick and M. Stephens (eds), *Industrial Performance, Education and the Economy in Victorian Britain* (Lewes: The Falmer Press, 1981).
P. W. S. Andrews and E. Brunner, *Capital Development in Steel: a Study of the United Steel Companies Ltd.* (Oxford: Basil Blackwell, 1952).

232 *Business Policies in the Making*

W. Ashworth, *An Economic History of England 1870–1939* (London: Methuen, 1960).

W. J. Baumol, *Business Behavior, Value and Growth* (New York: Macmillan, 1959).

Lady Bell, *At the Works* (Middlesbrough: privately printed, 1907).

J. S. Boswell, *Social and Business Enterprises* (London: Allen & Unwin, 1976).

'Hope, Inefficiency or Public Duty? The United Steel Companies and West Cumberland, 1918–39', *Business History* (January, 1980).

'The Informal Social Control of Business in Britain, 1880–1939', *Business History Review*, 1983.

M. Z. Brooke and H. L. Remmers, *The Strategy of the Multinational Enterprise* (London: Longmans, 1970).

D. L. Burn, *The Economic History of Steelmaking 1867–1939* (Cambridge: Cambridge University Press, 1940).

*The Steel Industry 1939–59* (Cambridge: Cambridge University Press, 1961).

T. H. Burnham and G. O. Hoskins, *Iron and Steel in Britain 1870–1930* (London: Allen & Unwin 1943).

N. K. Buxton, 'Efficiency and Organisation in Scotland's Iron and Steel Industry During the Inter-war Period', *Economic History Review* (February, 1976).

M. S. Birkett, 'The Iron and Steel Trades During the War', *Journal of the Royal Statistical Society*, LXXXIII (1920).

'The Iron and Steel Trades Since the War', *Journal of the Royal Statistical Society*, XCIII (1930).

F. Capie, 'The British Tariff and Industrial Protection in the 1930's', *Economic History Review* (August, 1978).

J. C. Carr and W. Taplin, *A History of the British Steel Industry* (Oxford: Basil Blackwell, 1962).

A. D. Chandler, *Strategy and Structure* (Cambridge, Mass.: MIT Press, 1962).

*The Visible Hand* (Cambridge, Mass.: Harvard University Press, 1977).

Sir H. Clay, *Lord Norman* (London: Macmillan, 1957).

J. Dean, *Managerial Economics* (New York: Prentice Hall, 1954).

D. Dougan, *The History of North East Shipbuilding* (London: Allen & Unwin, 1968).

A. Downs, *Inside Bureaucracy* (Boston, Mass.: Little, Brown, 1967).

A. R. Duncan, *A Memoir* (London: British Iron and Steel Federation, 1950).

C. Erickson, *British Industrialists: Steel and Hosiery 1850–1950* (Cambridge: Cambridge University Press, 1959).

J. K. Galbraith, *The New Industrial State* (London: Hamish Hamilton, 1967).

*Economics and the Public Purpose* (Boston, Mass.: Houghton Mifflin, 1973).

Sir. A. Grant, *Steel and Ships* (London: Michael Joseph, 1950).

H. M. Hallsworth, *An Industrial Survey of the North East Coast Area* (London: Board of Trade, 1932).

L. Hannah (ed.), *Management Strategy and Business Development* (London: Macmillan, 1976).

*The Rise of the Corporate Economy* (London: Methuen, 1976).

F. H. Hatch, *The Iron and Steel Industry of the U.K. under War Conditions* (Harrison and Sons, 1919).

C. J. Hawkins, *The Theory of the Firm* (London: Macmillan, 1973).

H. K. Hawson, *Sheffield, the Growth of a City 1893–1926* (Sheffield: J. W. Northend, 1968).

C. A. Hempstead (ed.), *Cleveland Iron and Steel* (London: British Steel Corporation, 1979).

E. P. Hexner, *The International Steel Cartel* (Chapel Hill, N. C.: University of Carolina Press, 1943).

R. Hidy *et al.*, *History of the Standard Oil Company*, 3 vols (New York: Harper and Bros., 1955–71).

J. Hodge, *Workman's Cottage to Windsor Castle* (London: Sampson Low, 1931).

J. R. Hume and M. S. Moss, *Beardmore: the History of a Scottish Industrial Giant* (London: Heinemann, 1979).

H. L. Hutchinson, *Tariff-making and Industrial Reconstruction: an Account of the Work of the Import Duties Advisory Committee 1932–39* (London: George Harrap, 1965).

'Ingot' (Sir Richard Clarke), *The Socialisation of Steel* (London: Gollancz, 1936).

Iron and Steel Trades Confederation, *Statement by the Executive Committee with Regard to the Situation in the Iron and Steel Industry* (1931).

R. J. Irving, *The North Eastern Railway Company 1870–1914* (Leicester: Leicester University Press, 1976).

J. Jewkes and A. Winterbottom, *An Industrial Survey of Cumberland and Furness* (Manchester: Victoria University, Dept. of Economics and Commerce, 1933).

R. Jones and O. Marriott, *Anatomy of a Merger, a History of G.E.C., English Electric and A.E.I.* (London: Jonathan Cape, 1971).

B. S. Keeling and A. E. Wright, *The Development of the Modern British Steel Industry* (London: Longmans, 1964).

M. W. Kirby, *The British Coalmining Industry 1870–1946* (London: Macmillan, 1977).

F. Knight, *Risk, Uncertainty and Profit*, with additional introductory essay (London: London School of Economics, 1946).

J. Y. Lancaster and D. R. Wattleworth, *The Iron and Steel Industry of West Cumberland, an Historical Survey* (London: British Steel Corporation, 1977).

John Lee (ed.), *Pitman's Dictionary of Industrial Administration*, 2 vols (London: Pitman, 1928–9).

S. Lloyd, *The Lloyds of Birmingham* (Birmingham: Cornish Bros Ltd, 1907).

A. F. Lucas, *Industrial Reconstruction and the Control of Competition: the British Experiments* (London: Longmans Green, 1937).

D. H. Macgregor *et al.*, 'Problems of Rationalisation', *Economic Journal*, XL (1930).

234    *Business Policies in the Making*

H. W. Macrosty, *The Trust Movement in British Industry* (London: Longmans Green, 1907)

D. N. McCloskey, *Economic Maturity and Entrepreneurial Decline: British Iron and Steel 1870–1913* (Cambridge, Mass.: 1973)

R. L. Marris, *The Economic Theory of Managerial Capitalism* (London: Macmillan, 1964).

R. L. Marris and A. Wood (eds), *The Corporate Economy* (London: Macmillan, 1971).

A. Marshall, *Principles of Economics*, 1st edn (London: Macmillan, 1890).

'Some Aspects of Competition', Address to Economic Science and Statistics Section, British Association, Leeds, 1890.

K. Middlemas, *The Politics of Industrial Society* (London: André Deutsch, 1979).

W. E. Minchinton, *The British Tinplate Industry, a History* (Oxford: Clarendon Press, 1957).

E. V. Morgan, *Studies in British Financial Policy, 1914–25* (London: Macmillan, 1953).

Ministry of Munitions, *History*

A. Nevins and F. E. Hill, *Ford, the Times, the Man and the Company* (New York: Charles Scribner's Sons, 1954).

G. A. North, *Tees-side's Economic Heritage* (Middlesbrough: Cleveland County Council, 1975).

R. J. Overy, *William Morris, Viscount Nuffield* (London: Europa, 1976).

P. L. Payne, 'The Emergence of the Large-scale Company in Great Britain 1870–1914', *Economic History Review*, 20 (1967).

P. L. Payne, 'Rationality and Personality: a Study of Mergers in the Scottish Iron and Steel Industry 1916–36', *Business History* (July, 1977).

*Colvilles and the Scottish Steel Industry* (Oxford: Clarendon Press, 1979).

R. Peddie, *The United Steel Companies* (Sheffield: The United Steel Companies, 1967).

E. Penrose, 'Biological Analogies in the Theory of the Firm', *American Economic Review* (December, 1952).

*Theory of the Growth of the Firm*, 2nd edn (Oxford: Basil Blackwell, 1979).

P. Poerschke, 'Kostenerfassung, Verbands-und Absatzorganisation in der Stahlrohren-Industrie' (Freiburg Ph.D.; 1935).

S. Pollard, *A History of Labour in Sheffield* (Liverpool: Liverpool University Press, 1959).

*The Development of the British Economy 1914–50*, 2nd edn (London: Arnold, 1969).

A. Pugh, *Men of Steel by One of Them: a Chronicle of 88 Years of Trade Unionism in the British Iron and Steel Industry* (London: Iron and Steel Trades Confederation, 1951).

W. J. Reader, *Architect of Air Power: the Life of 1st Viscount Weir of Eastwood 1877–1959* (London: Collins, 1968).

*I.C.I., a History*, 2 vols. (London: Oxford University Press, 1971, 1975).

J. M. Rees, *Trusts in British Industry, 1914–21* (London: P. S. King and Son, 1922).

J. M. Reid, *James Lithgow* (London: Hutchinson, 1964).

H. W. Richardson and J. M. Bass, 'The Profitability of Consett Iron Company Before 1914', *Business History* (July, 1965).

M. W. Ridley, Viscount Ridley, *Development of the Iron and Steel Industry in North West Durham* (Newcastle-upon-Tyne: King's College, 1961).

W. Robertson, *Middlesbrough's Effort in the Great War* (Middlesbrough: Jordison and Co, 1922).

R. S. Sayers, *The Bank of England 1891–1944*, 3 vols (Cambridge: Cambridge University Press, 1976).

F. Scopes, *The Development of Corby Works* (Stewarts and Lloyds, 1968).

J. D. Scott, *Vickers, A History* (London: Weidenfeld & Nicolson, 1963).

J. Schumpeter, *Capitalism, Socialism and Democracy*, 5th edn (London: Allen & Unwin, 1976).

A. Singh and G. Whittington, *Growth, Profitability and Valuation* (Cambridge: Cambridge University Press, 1968).

A. Slaven, 'A Shipyard in Depression: John Browns of Clydebank 1919–38', *Business History* (July, 1967).

P. Stubley, 'The Churches and the Iron and Steel Industry in Middlesbrough 1890–1914' (Durham M. A., 1979).

The Times, *Books of New Issues* (London: The Times, 1918–24).

S. Tolliday, 'Industry, Finance and the State: an Analysis of the British Steel Industry in the Inter-war years' (Cambridge Ph.D., 1979).

'Tariffs and Steel, 1916–33, the Politics of Industrial Decline' (unpublished paper read at London School of Economics seminar, 1980).

J. Vaizey, *The History of British Steel* (London: Weidenfeld & Nicolson, 1974).

G. R. Walshaw and C. A. J. Behrendt, *The History of Appleby-Frodingham* (Appleby-Frodingham Steel Company, 1950).

E. Wilkinson, *The Town that was Murdered: the Life-story of Jarrow* (London: Gollancz, 1939).

O. E. Williamson, *Markets and Hierarchies* (New York and London: The Free Press, 1975).

W. G. Willis, *South Durham Steel and Iron Company Ltd.* (South Durham Steel and Iron Company, 1969).

A. S. Wilson, 'The Consett Iron Company Ltd., a Case Study in Victorian Business History' (Durham Ph.D., 1973).

C. Wilson, *Unilever*, 3 vols (London: Cassell, 1954–70).

*A Man and His Times: a Memoir of Sir Ellis Hunter* (London: Newman Neame, c. 1968).

O. Wood, 'The Development of the Coal-mining, Iron and Shipbuilding Industries in Cumberland 1950–1914' (Liverpool Ph.D. 1952).

# Index

## 238 *Business Policies in the Making*

Dorman Long *cont.*
Ellis Hunter assumes power 157–9;
expansion of 36–42; expansionist
projects 79, 81; extracts from
internal documents 103–6; financial
errors 82–3; financial weakness 103,
106; Francis Samuelson, director 30,
78, 86n; growth and efficiency
159–63; inefficiency of 79, 118;
labour policies 166–7, 168; lack of
funds 79; mergers 36, 38; move
towards efficiency 156, 158–9;
munitions manufacture 38–9;
overseas activities 36, 81, 82, 209;
philanthropy and 176–7, 218;
political-public passivity 185–8;
pollution and 177–8; profits 37, 38–9,
62; rationalisation 118; Redcar
scheme 39, 40, 41, 42, 57, 79, 213;
social action and 202–5, 218–22;
Sydney Bridge contract 81, 110, 118;
technology and 80; top management
103, 106–7, 108–10; vertical
integration 36; vertical
rationalisation 35–6; weakness of
120–1; welfare policies 167; *see also*
Sir Arthur Dorman, Ellis Hunter
and Charles Mitchell
Downs, Anthony 12, 18n
Duncan, Sir Andrew 109, 117, 132,
134, 136, 183, 187, 188, 194, 204

economic cycles 2, 11
economic determinism 15, 20, 63
economic historians and steel industry
28
efficiency 8, 13–15, 63, 215–18, 226;
definition of 9–10; growth and 11–12,
159–63; growth-e. conflict 11–12, 28;
growth-e. trade-offs 31; iron and
steel industry 20; lack of e. in
Dorman Long 79; lack of e. in USC
75–6; mergers and 32; period 1921–8
83–5; social action and 11, 219;
Stewarts and Lloyds and 66, 69, 70,
83, 93, 95, 154, 156; ways of
improving e. 30–1; *see also*
individual companies
Ellis, J. V. 72, 73, 148
Ennis, Laurence 78–9, 110, 158, 161,
188
entrepreneurs 14, 42
environment 33, 177–9
error 7–8, 20, 209, 213; typology
213–14; e. of commission 57; e. of

omission 57–8; within Dorman Long
82–3; within USC 76

Federation of British Industry 196
France 25
Frodingham Iron and Steel Company
37, 45, 46, 64, 73, 75, 96, 133, 145,
152
Furness family 115, 118

Galbraith, J. K. 9
Gardner, Charles Bruce 117, 136, 192,
199
General Strike, the 62, 168
Germany 25, 123, 124
Goodenough, F. C. 103, 105–6, 117
government 26, 27, 41, 43, 183, 189,
191, 192, 193, 195, 198, 202, 203
Greenwood, Lord (Harmar) 158
growth 8, 13–15, 17–18, 215–18, 222–6;
definition of 9; efficiency and 11–12,
159–63; g. through mergers 32;
g.-efficiency conflict 11–12, 28;
g.-efficiency trade-offs 31; iron and
steel industry 20; social action and
11, 219; Stewarts and Lloyds and 65,
84, 93, 95, 121, 128, 129–30; USC
and 130–7; *see also* individual
companies
Guedalla, F. E. 72, 76, 96, 98

Hatry, Clarence 143–4
Head, C. A. 78
Henderson, James 96, 148
Hickman, Brigadier General 63
Highton, Langton 149
Hilton, Robert 98, 100–2, 110, 111, 112,
136, 141, 142, 144, 148, 149, 150–1,
152, 153, 171, 190–3, 195, 210, 211,
217, 222
Howard, Henry 52, 53, 63, 90
Howard, Joseph Jr 53, 64, 155
Hunter, Ellis 157–9, 185, 187, 188, 210,
217, 222

Imports Duties Advisory Committee
(IDAC) 183, 188, 193, 198, 201–2;
National Committee 187, 192–3, 199
indeterminacy, managerial hiatus 16,
17
India 70, 81
information systems 30, 68–9, 149
investment 11, 29, 31, 63, 131–2
iron and steel industry 20–1; capital
investment in 23; cartelisation
within 182; cartels and 23, 26, 27;